Ultrasonic Measurements and Technologies

Sensor Physics and Technology Series
Series editors:
Professor K. T. V. Grattan
Centre for Measurement, Instrumentation and Applied Physics
City University
London, UK

Dr A. Augousti
School of Applied Physics
Kingston University
Kingston-upon-Thames, UK

The *Sensors Physics and Technology Series* aims to bring together in a single series the most important developments in the rapidly-changing area of sensor technology and applications. It will present a snapshot of the range of effort which is being invested internationally in the development of novel types of sensors. New workers in the area of sensor technology will also be catered for with an introduction to the subject through the provision of tutorial guides. Volumes may be sensor technology or applications oriented, and will present recent results from the cutting edge of research in a compact monograph format.

Topics covered will include:

- optical sensors: free-space sensors
- optical sensors: guided wave sensors
- solid state sensors
- biosensors
- microwave sensors
- ultrasonic sensors
- process tomography
- control of networked sensors system control and data aquisition
- medical instrumentation
- infrared sensors
- chemical and biochemical sensing
- environmental sensing
- industrial applications

Titles available:

1. Biosensors
   Tran Minh Cahn

2. Fiber Optic Fluorescence Thermometry
   K. T. V. Grattan and Z. Y. Zhang

3. Silicon Sensors and Circuits
   F. Wolffenbuttel

Ultrasonic Measurements and Technologies

Štefan Kočiš
Faculty of Electrical Engineering and Information Technology
Slovak Technical University
Bratislava, Slovak Republic

and

Zdenko Figura
SONO Electric
Nové Mesto nad Váhom
Slovak Republic

CHAPMAN & HALL
London · Weinheim · New York · Tokyo · Melbourne · Madras

Published by Chapman & Hall, 2–6 Boundary Row, London SE1 8HN, UK

Chapman & Hall, 2–6 Boundary Row, London SE1 8HN, UK

Chapman & Hall GmbH, Pappelallee 3, 69469 Weinheim, Germany

Chapman & Hall USA, 115 Fifth Avenue, New York, NY 10003, USA

Chapman & Hall Japan, ITP–Japan, Kyowa Building, 3F, 2-2-1 Hirakawacho, Chiyoda-ku, Tokyo 102, Japan

Chapman & Hall Australia, 102 Dodds Street, South Melbourne, Victoria 3205, Australia

Chapman & Hall India, R. Seshadri, 32 Second Main Road, CIT East, Madras 600 035, India

---

Published in co-edition with Ister Science Limited
Ister Science Limited, Staromestská 6, Bratislava, Slovak Republic
Typeset by Forma Ltd., Bratislava, Slovak Republic

First edition 1996

Printed in Great Britain by St Edmundsbury Press, Bury St Edmunds, Suffolk

ISBN 0 412 63850 9 (Chapman & Hall)
ISBN 80-88683-09-2 (Ister Science)

A catalogue record for this book is available from the British Library

∞ Printed on permanent acid-free text paper, manufactured in accordance with ANSI/NISO Z39.48-1992 (Permanence of Paper)

Contents

Preface

An impulse for writing this book has originated from the effort to summarize and publicise the acquired results of a research team at the Department of Automation of the Faculty of Electrical Engineering and Informatics, Slovak Technical University in Bratislava. The research team has been involved for a long time with control problems for machine production mechanisms and, in recent (approximately 15) years, its effort was aimed mostly at the control of electrical servosystems of robots. Within this scope, the members of the authors' staff solved the State Research Task Ultrasonic sensing of the position of a robot hand, which was coordinated by the Institute of Technical Cybernetics of the Slovak Academy of Sciences in Bratislava.

The problem was solved in a complex way, i.e. from a conceptual design of the measurement, through the measurement and evaluation system, up to connection to the control system of a robot. Compensation of the atmospheric influence on the precision of measurement, as well as on the electroacoustical transducers, were important parts of the solution. The solution was aimed at using the ultrasonic pulse method which enables the measurement of absolute 3D position coordinates, contrary to the relative position measurements by the incremental pick-ups which are standard robotic equipment.

The selected configuration of transducer arrangements and the coordinate calculation method enables us to compensate fully for the influence due to variations of parameters of the ambient medium, such as temperature, humidity, etc., except for air flow which has a vector character. However, in the environments where robots work, the air flow has a considerable influence on the precision of a measurement. Therefore, correction of this effect was attempted by introducing an auxiliary measurement of the air flow vector. The measurement method was selected with respect to full utilization of the given configuration of transducers, together with the appropriate electronic circuits of individual measurement channels. On mutual exchange of two measuring channels (with reciprocal transducers), another phase of measurement is carried out (in the first phase, the coordinates are measured). After evaluation of that second phase, the individual terms of the flow vector are calculated, using particular algorithms. Then correction of the coordinates measured in the first phase follows.

As a by-product of solving the task of the air-flow measurement, a possibility of its utilization as a self-standing, separate instrument for air-flow measurements, (e.g. the wind, for meteorological purposes), can be considered. By extension of the calculation algorithm, it is also possible to calculate the air temperature from the measured readings. For the purpose of this book, this problem was elaborated by Pavol Bystrian-sky, M.S., PhD, in parts 6.2 and 7.3.

An important problem of electroacoustical transducers for generating spherical pulse waves, which is necessary for picking up the spatial coordinates, was described by Miroslav Toman, M.S., PhD in parts 3.1 and 3.4. The problems of control and evaluation of the measurement process for the purpose of testing the trajectories of the robots motion was described by Július Oravec M.S. in part 7.2.

In order to extend the problems in the book, the authors' team was joined by Zdenko Figura, M.S., PhD, a former long-term chief of the Ultrasonic Division of the Research Institute for Mechanization and Automation Control in Nové Mesto nad Váhom, who at present is a manager of SONO Electric Ltd. He deals with ultrasonic sensors, especially with applications in machinery, electrotechnical, chemical, and civil engineering technology. He has put together Chapter 2 on the physics of ultrasonic waves; furthermore, he elaborated parts 3.2, and 3.3, of Chapter 3 (on ultrasonic transducers) and parts 4.2, 4.4, 4.5 and 4.6 of Chapter 4 (on ultrasonic techniques). He also wrote chapter 5 on non-destructive testing (NDT) completely and he contributed to Chapter 7 in parts 7.4 to 7.7 (examples of ultrasound applications).

Other parts of the book were written by the leader of the authors' team, Assoc. Prof. Štefan Kočiš, M.S., PhD, who wishes to express his sincere thanks to the co-authors for their contribution to the manuscript. The leader of the authors' team especially wants to thank the authors' employers for enabling them to acquire the information necessary for writing the manuscript.

1 Introduction

At present, the term ultrasound has acquired a wide meaning, involving whole fields of physics, industrial technology, information and measurement technology, medicine, and biology. The term ultrasound means a branch of acoustics which uses the acoustic band with frequencies above the audible limit. The upper frequency limit is being constantly increased and, at present, it shifts towards the region of hypersound.

Many principles of classical acoustics are usable in the ultrasonic area too. However, considerable development of ultrasound technology has occurred especially after the discovery of phenomena characteristic of the region of ultrasonic frequencies .

The 19th century prepared suitable conditions for the rise of ultrasonic technology. However, only in the twentieth century did more significant development begin. It is said that ultrasound was born in 1917, with the invention of sonar. Some more important industrial applications appeared during World War 2, especially in the branch of non-destructive testing (NDT) and soldering. After the war, ultrasound has spread into many fields and at present it represents a huge field involving science, industry, and medicine.

In recent decades acoustoelectronics and acoustooptics have appeared as independent areas. Acoustoelectronics is used for the processing of electronic signals and their conversion to ultrasonic ones, e.g. in delay lines, filters and semiconductor elements, which exploit surface hypersound waves. Acoustooptics deals with treatment of light signals by ultrasound. Acoustic holography, which enables imaging (scanning, displaying, projection) of objects in opaque media is a rapidly developing method.

Apart from the above mentioned areas of application, there exists an innumerable variety of industrial applications. Let us mention, at least, a few of these: measurement of mechanical quantities, like position, distance, thickness, velocity, level height and others, like flow, temperature, pressure, vibration, elasticity, density, composition, etc.

The applications mentioned above belong to the area of low-power (signalling) ultrasound. A very broad field of application is in the region of high-power ultrasound. This field is not covered by this book. For the sake of completeness, however, let us mention at least some of these applications here: cleaning, surfacing and machine-working, welding, sputtering (atomization), emulsification, influencing chemical reactions, therapy, surgery, and others.

In respect to a broad spectrum of applications, it is neither possible nor suitable to cover all applications of low-power ultrasound techniques in this book. The less important and less developed ones have been omitted as well as those, however important, which are not of a sensing character (e.g. delay lines, filters, medical, biological, and research applications).

A short chapter about the fundamental physical principles of ultrasound propagation has been included in the book. It should contribute to a better understanding of the principle of operation of instruments and equipment, and especially of the mutual interaction between an ultrasonic transducer and the measured parameter. This can be important for determining the choice of a suitable type of instrument. The selection of a standard instrument, for an application beyond the scope of its limitations, can lead to a loss of time and money. And moreover these limitations may not be known even to the seller at purchase.

For the successful application of an ultrasonic instrument, not only is a knowledge and understanding of the interaction of ultrasound with the examined parameter necessary, but also the securing of efficient sound transfer between a transducer and the material under investigation. In most applications, built-in (internal) probes are used more often than built-on (external) ones. One particular reason for this is that internal probes are easier to design with respect to compliance with the measured parameters, and, as a rule, they are more resistant to the influence of external effects. External probes often require considerable experience and skill for the attainment of acoustical matching. Therefore the user should know the conditions required for the proper operation of the probe.

In some applications, especially in the cases of measuring mechanical quantities, several alternative solutions can exist, for example, based on optical, electromagnetic, capacitive and other principles. Therefore the advantages and drawbacks should be compared from different points of view. The ultrasonic solution also has its advantages and drawbacks which, in a particular application, may have different weights. As advantages, the relative simplicity of generating and receiving ultrasonic waves can be considered as well as the compactness of transducers, their insensitivity to the influence of the surrounding medium, high reliability and a relatively low price. On the other hand, it should be remembered (especially in regard of gaseous media) that the velocity of propagation of ultrasonic waves is influenced by the temperature and humidity of the working medium in which the ultrasonic wave propagates, as well as by some kinds of acoustic interference. Therefore, the instruments often require auxiliary correction and regulation elements.

For these and other reasons, the electronic circuits of more precise instruments are usually more complicated. This may explain the considerable time lag between the discovery of the principles of ultrasonic measurements and their realization which was made possible by the rapid increase in the development of electronics and computer technology.

Perhaps the delay in the development of exploitation of ultrasound, in contrast with optics, may also be explained by the fact that ultrasonic waves are not perceived by any of the human senses, while using sight, a human being is able to perceive and recognize objects with high precision and sensitivity. This is why man had to develop ultrasonic sight with great difficulty, and slowly. In 1794, the Italian Lazzaro Spallanzani tried to explain the ability of bats to fly about an obstacle in full darkness, by utilization of their hearing. But this ability of bats to exploit echolocation for identifying obstacles was proven definitely only as late as 1945. At present, the exploitation of the echolocation principle has become highly advanced. A proof of this are the many kinds of instruments for defectoscopy, sonography, and so on.

From the point of view of ultrasonic seeing, it is interesting to compare the properties of light and ultrasonic waves. Light propagates well through many gases, slightly worse through liquids, and only through a limited number of solids. Ultrasound passes (more or less) through all materials. Light passes through vacuum without losses, whereas sound does not propagate at all. There is a physical reason: sound propagates by wave motion of massive particles. Therefore, the use of ultrasonic measuring methods is reasonable in the cases where a non-transparent material may be penetrated by this wave motion, and examined internally.

It should be noted that ultrasonic seeing has a low resolution compared with light and it also applies to much shorter ranges. This is caused by certain obstacles to the propagation of ultrasound, namely by dispersion and attenuation (a more detailed explanation is given in Chapter 2).

These obstacles are slowly but continuously being overcome by the introduction of integrated matrix sensors produced by microelectronic technology. In this way, an ultrasonic beam can be focused and deflected, so that the entire space in question can be scanned.

By the use of ultrasound, the distance to the surface of the observed object can also be determined. It is performed by using the above mentioned echolocation method. In this method, the time of propagation of an ultrasonic pulse is measured, from the instant of its emission to its return onto the active surface of the transducer, after reflection from the surface of the observed object. In this case, the transmitter and the receiver of the pulse can be the same. These reciprocal transducers (capacitive, piezoelectric, magnetostrictive ones) are used most often and they have many advantages.

In selecting the contents of this book, a certain compromise had to be made. Some problems are described in detail, others are only briefly outlined. It also relates to the professional orientation and practical experience of individual members of the authors' team. This team stands by the opinion that a more detailed description of some methods and instruments can help the reader to also grasp an understanding of the applications which are described more briefly. Moreover, it enables him to judge the possibility of application of ultrasonic techniques and technology for his actual needs, on a more qualified level.

2 Ultrasonic waves

In an elastic material medium, mechanical waves of various frequencies can propagate. For frequencies above the audible band, these are called ultrasonic waves. They differ from sound waves in frequency.

An oscillatory motion can be excited by a change from an equilibrium state of a mass particle, by acting on it with a certain force. If the oscillatory motion is repeated during a certain time interval with a period T, it is called a periodic or harmonic motion. A reciprocal value of the period T is the frequency, f, which is defined as the number of oscillations of the mass particle per unit time. The unit for frequency is Hz $[s^{-1}]$

$$f = 1/T \quad [\mathrm{Hz, s}].$$

An oscillating mass particle always forms a system with the surrounding medium in which the particles are bound together by elastic bonds. An oscillation or a wave motion of the elastic medium is caused by a displacement and an oscillatory motion of a particle group which act on their neighbouring particles through elastic bonds. The disturbance propagates in the medium as a wave motion with a certain velocity, c, which depends on the physical properties of the medium. Thus, the wave motion is characterized by two important features: by a propagation, and by a transmission of energy, without a transmission of mass.

If the direction of individual particle oscillations corresponds at all points of the medium with the direction of a wave propagation, the wave motion is longitudinal. If these two directions are mutually perpendicular, the wave motion is transverse. The existence of these two types of wave motion depends on the physical state of the medium, i.e. solid, liquid or gaseous.

If the oscillating particles follow the sinusoidal law of motion, then the oscillations can be described by the following wave equation

$$\frac{\partial^2 a}{\partial t^2} = c\frac{\partial^2 a}{\partial x^2} \qquad (2.1).$$

A particular solution for a planar wave is the equation for harmonic motion

$$a = A\sin\omega\left(t - \frac{x}{c}\right) \qquad (2.2)$$

where

A	- amplitude of displacement [m]
$\omega = 2\pi f$	- circular (or angular) frequency [rad.s^{-1}]
c	- velocity of wave propagation [m.s^{-1}]
x	- position coordinate [m]
$\omega x/c$	- phase angle φ [rad]
a	- displacement [m].

The velocity of the particle oscillations around the equilibrium position, the so called acoustic velocity, v, is given by the time derivative of the instantaneous displacement, a

$$v = \frac{da}{dt} = \omega A\cos(\omega t - \varphi) \quad [\mathrm{m.s^{-1}}] \qquad (2.3).$$

If a sound wave propagates in a medium with a velocity c, then the wavelength λ, the period, T, and the frequency, f, are related together by the formula

$$\lambda = cT = \frac{c}{f} \qquad (2.4).$$

On transition of the ultrasonic wave through a medium, an acoustic pressure, p, is created. The pressure is related to the acoustic velocity, v, by the formula

$$p = zv \qquad (2.5),$$

where z is the specific acoustic impedance.

The equation (2.5) is an acoustic analogy of Ohm's law in electrical engineering. The pressure, p, corresponds to an electric voltage, the acoustic impedance, z, to an electric impedance. The acoustic impedance, z, is generally a complex value, similarly to an electric impedance, because of the possible phase shift between the acoustic pressure and the acoustic velocity. It has considerable significance in the case of a travelling wave in an infinite medium where the acoustic impedance has a real value, and is called the wave resistance, ρc.

If an ultrasonic wave, with an acoustic velocity given by equation (2.3), propagates through a medium with a wave resistance, ρc, it gives rise to an acoustic pressure

$$p = \omega A \rho c \,.\, \cos\omega \left(t - \frac{x}{c} \right) \quad [\mathrm{Pa}] \qquad (2.6)$$

where

$\omega A \rho c = P$ - amplitude of acoustic pressure [Pa]
$\omega A = V$ - amplitude of acoustic velocity [m.s^{-1}].

As in electrical engineering, the effective (root-mean-square) variables

$$p_{ef} = \frac{p}{\sqrt{2}} \,, \qquad v_{ef} = \frac{v}{\sqrt{2}}$$

can be introduced. Using these variables, the intensity of an ultrasonic wave can be defined as

$$I = p_{ef} \cdot v_{ef} = \frac{1}{2} p \,.\, v = \frac{1}{2} \frac{p^2}{\rho\, c} \quad [\mathrm{W.m^{-2}}] \qquad (2.7)$$

Thus, the intensity, I, is a power passing through a unit area, s, and is proportional to the square of the acoustic pressure, p. The power of the ultrasonic wave is expressed as

$$N = p_{ef} \cdot v_{ef} \cdot s \quad [\mathrm{W}] \qquad (2.8).$$

2.1 Propagation of ultrasound in solids, liquids and gases

All wave processes are characterized by a wavelength, λ, an amplitude of displacement, A, and a velocity of propagation, c. Most wave processes are characterized by additional parameters - a period of oscillation, T, a frequency, f, and an angular frequency, ω. These quantities are related together by the following formulae:

$$\lambda = cT \quad [\text{m, m.s}^{-1}\text{, s}]$$
$$f = 1/T \quad [\text{Hz, s}]$$
$$\omega = 2\pi f \quad [\text{rad.s}^{-1}\text{, Hz}]$$

In solids, various kinds of waves can propagate with different velocities.

In an unlimited medium i.e. in a medium whose transverse dimensions are considerably larger than the wavelength, the velocity of propagation is given by the formula

$$c_k = \sqrt{\frac{E_p}{\rho}} \cdot \sqrt{\frac{1-\mu}{(1+\mu)(1-2\mu)}} \quad [\text{m.s}^{-1}] \qquad (2.9)$$

where

E_p - modulus of elasticity in tension (Young's modulus) [Pa]
ρ - density [kg.m^{-3}]
μ - Poisson's ratio.

Let us consider a medium whose one dimension is limited in the direction perpendicular to the direction of wave propagation (a plate). If for the thickness d, the condition $d \ll \lambda$ is true, then the propagation velocity, c_L, of the longitudinal waves is

$$c_L = \sqrt{\frac{E_p}{\rho} \cdot \frac{1}{1-\mu^2}} \quad [\text{m.s}^{-1}] \qquad (2.10).$$

Let us consider another case, when two dimensions of a solid material medium are limited in the directions x and y, both perpendicular to the direction of the ultrasound propagation (e.g. a bar). If the conditions $x \ll \lambda$, $y \ll \lambda$ are true, then the velocity of propagation of the longitudinal wave motion is

$$c_L = \sqrt{\frac{E_p}{\rho}} \quad [\mathrm{m.s^{-1}}] \qquad (2.11).$$

The propagation velocity of transverse waves in a solid medium can be expressed as

$$c_T = \sqrt{\frac{G_s}{\rho}} = \sqrt{\frac{E_p}{\rho} \cdot \frac{1}{2(1+\mu)}} \quad [\mathrm{m.s^{-1}}] \qquad (2.12)$$

where G_s is a modulus of transverse elasticity [Pa].

The velocity of propagation as well as the wavelength at a given frequency are different for different types of waves. The velocity of transverse waves and consequently, the wavelength too, are approximately half the values for longitudinal waves provided the medium and frequency stay unchanged. Their ratio depends only on Poisson's number, μ:

$$\frac{c_T}{c_L} = \frac{1-2\mu}{2(1-\mu)} \cong \frac{1}{2} \qquad (2.13).$$

The values of the velocity of ultrasonic wave propagation, and other relevant physical constants of solids which may be important for ultrasonic technology, are given in Table 2.1.

The velocity of surface waves, c_R, is always less by 5 to 10 % than the velocity of transverse waves, c_T. An approximate expression, as given in Ref. [2.1] is

$$c_R \cong c_T\left(\frac{0.87+1.12\mu}{1-\mu}\right) \quad [\mathrm{m.s^{-1}}] \qquad (2.14).$$

In pure liquids, longitudinal ultrasonic wave motion can propagate. Its velocity is given by

$$c = \sqrt{\frac{1}{\beta_{ad}\rho}} = \sqrt{\varkappa \frac{1}{\beta_{is}\rho}} \quad [\mathrm{m.s^{-1}}] \qquad (2.15)$$

where

β_{ad} - adiabatic coefficient of compressibility [Pa^{-1}]

β_{is} - isothermal coefficient of compressibility [Pa^{-1}].

TABLE 2.1

Constants of some solid media at a temperature 20 °C

medium	density $.10^3$ [$kg.m^{-3}$]	Poisson's number μ	propagation velocity c [$m.s^{-1}$]		wave resistance $Z_e = \rho.C_L.10^6$ [$Pa.s.m^{-1}$]
			longitudinal	transverse	
metals					
tin	7.30	0.33	3320	1670	24.2
aluminium	2.70	0.34	6320	3080	17.0
magnesium	1.73	0.30	5780	3050	10.0
cast iron	7.20	-	3500 – 5600	2200 – 3200	25 – 40
brass	8.10	0.35	3830	2123	31.0
nickel	8.80	0.31	3830	2960	49.5
steel	7.80	0.28	5900 – 6000	3200	46.3
non-metals					
ice	1	0.33	3980	1990	3.98
organic glass	1.18	0.35	2670	1121	3.2
polystyrene	1.06	0.32	2350	1120	2.40
porcelain	2.40	-	5300 – 5500	-	13
quartz glass	2.60	0.17	5570	3515	14.5
teflon	2.2	-	1350	-	3
rocks					
basalt	2.72	0.3	5930	3140	16.2
slate	2.74	0.27	6300	3610	17.8
marble	2.66	0.30	6150	3260	16.4
granite	2.62	0.18	4450	2780	11.6

The velocity of an ultrasonic wave in liquids is temperature dependent. For practical use, this dependence can be expressed for most liquids by the linear empirical relation

$$c = c_o + \gamma\theta \qquad [m.s^{-1}] \qquad (2.16)$$

where

$\theta = \Theta - \Theta_o$ - temperature [°C]

Θ, Θ_o - temperature, absolute temperature [K]
($\Theta_o = 273.15\ K = 0\ °C$)

c_o - velocity at the initial temperature Θ_o [$m.s^{-1}$]

γ - absolute temperature coefficient [$m.s^{-1}.K^{-1}$].

Constants of some liquids at the temperature 20 °C are given in Table 2.2.

TABLE 2.2

Constants of some liquids at a temperature 20 °C

liquid	density.10^3 [$kg.m^{-3}$]	propagation velocity c [$m.s^{-1}$]	absolute temperature coefficient γ [$m.s^{-1}.K^{-1}$]	wave resistance $Z_o=\rho.C.10^4$ [$Pa.s.m^{-1}$]
acetone	0.792	1192	-5.5	94
ethanol	0.789	1180	-3.6	93
petrol	-	1162	-4.14	-
glycerol	1.261	1923	-1.8	242
chloroform	1.489	1005	-	149
methanol	0.792	1128	-3.3	89
olive oil	0.905	1405	-	127
paraffin oil	0.835	1444	-	121
linseed oil	0.922	1923	-	165
toluene	0.866	1328	-	65
tetrachloromethane	1.595	938	-3	150
water	0.997	1483	+2.5	148
transformer oil	0.9	1400	-	130

Generally speaking, the velocity of ultrasonic waves in liquids (except water) decreases with increasing temperature. For water, the velocity of ultrasound propagation increases with increasing temperature until it reaches a maximum at 74 °C and then decreases (Fig. 2.1).

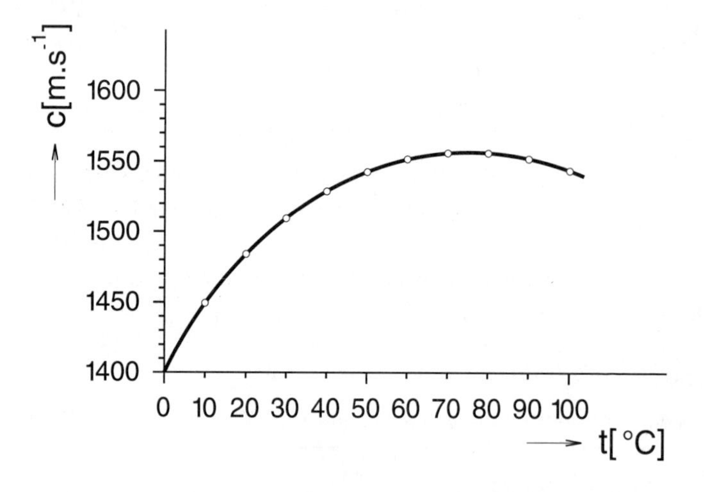

Fig. 2.1 ***Temperature dependence of ultrasound propagation in water***

The influence of hydrostatic pressure on the propagation velocity is negligible. The change is about 0.1 % per MPa.

The velocity of propagation in solutions and suspensions depends on the concentration of individual components. In aqueous salt solutions, the velocity of propagation increases linearly with the concentration. In aqueous solutions of some acids, a linear dependence of the velocity of propagation is maintained up to some limiting concentration and, above this point, with further increase of concentration, it decreases in an approximately linear manner.

In suspensions and colloidal solutions, which contain small particles, the velocity of ultrasound propagation depends on their concentration as well as on their size, especially if the particle size is less than the wavelength [2.2].

In an ideal gas, the velocity of ultrasound propagation can be expressed by more, mutually related formulae. For example, it is often expressed as

$$c = \sqrt{\varkappa \frac{p_a}{\rho}} \quad [m.s^{-1}] \tag{2.17}$$

where p_a - total atmospheric pressure (~10^5 Pa)
$\varkappa = c_p/c_v$ - the ratio of specific heats at constant pressure and constant volume (Poisson's constant)
ρ - gas density (e.g. 1.189 $kg.m^{-3}$ for air).

As in liquids, the propagation velocity can also be expressed by adiabatic and isothermal coefficients of compressibility, β_{ab}, and β_{is}, and Poisson's constant, $\varkappa$

$$c = \sqrt{\frac{1}{\beta_{ad}\rho}} = \sqrt{\varkappa \frac{1}{\beta_{is}\rho}} \quad [m.s^{-1}] \tag{2.18}$$

where $\varkappa = \frac{\beta_{is}}{\beta_{ad}} = \frac{c_p}{c_v}$.

The propagation velocity of ultrasound depends also on temperature. For small variations, the relationship is

$$c = c_o(1 + \gamma_r\theta) = c_o + \gamma\theta \quad [m.s^{-1}] \tag{2.19}$$

where

$c_o = \sqrt{\gamma \frac{p_a}{\rho_0}}$ - propagation velocity of ultrasonic waves in gas at 0 °C
γ_r - relative temperature coefficient $[K^{-1}]$
$\gamma = c_0\gamma_r$ - absolute temperature coefficient $[m.s^{-1}.K^{-1}]$
$\theta = \Theta - \Theta_o$ - temperature
Θ - absolute temperature [K].

Constants of some gases at a temperature 0 °C are given in Table 2.3.

With increasing humidity, the propagation velocity increases approximately linearly. For example, at the level of humidity of 50 %, the velocity increases by 1 %, at 100 % by 2 %.

TABLE 2.3

Constants of some gases at a temperature 0 °C

gas	symbol	propagation velocity c [m.s^{-1}]	temperature coefficient γ [m.s^{-1}.K^{-1}]	ratio of specific heats κ
argon	A	319	0.56	1.668
helium	He	965	0.8	1.66
carbon dioxide	CO_2	259	0.4	1.299
carbon monoxide	CO	338	0.6	1.4
air		331	0.61	1.402
neon	Ne	435	0.8	–
oxygen	O_2	316	0.56	1.396
sulphur dioxide	SO_2	213	0.47	1.29
nitrogen	N_2	334	0.6	1.4
hydrogen	H_2	1284	2.2	1.408

In real gases (especially the polyatomic ones), the propagation velocity depends on frequency. This was not accounted for in the foregoing expressions for ideal gases. This frequency dependence is explained by acoustic dispersion which is influenced by various physical parameters of the medium such as, for instance, internal friction in the gas, relaxation phenomena and absorption stimulated by resonance phenomena. The frequency dependence of the propagation velocity in gases is influenced for the most part by thermal relaxation, characterized by a dclay, i.c. by the relaxation time, necessary for an exchange of energy between the external and the internal degrees of freedom.

2.2 Reflection and transmission of an ultrasonic wave

The propagation of ultrasonic waves is influenced by any boundary surface (interface). At the boundary, the wave is reflected, and in the case of oblique incidence, refracted as well. If the boundary surface is large enough, it is possible to apply the laws of reflection and refraction to the incident ultrasonic beam, as known from geometrical optics.

A perpendicular incidence of an ultrasonic wave onto the boundary of two media is illustrated in Fig. 2.2. Let the wave resistance of a medium A be Z_{01}, and the wave intensity be I_{a1}. This ultrasonic wave penetrates a medium B with a wave resistance Z_{02}, and with an intensity I_{b1}. Then the intensity of the reflected wave motion is

$$I_{a2} = I_{a1} - I_{b1} \ .$$

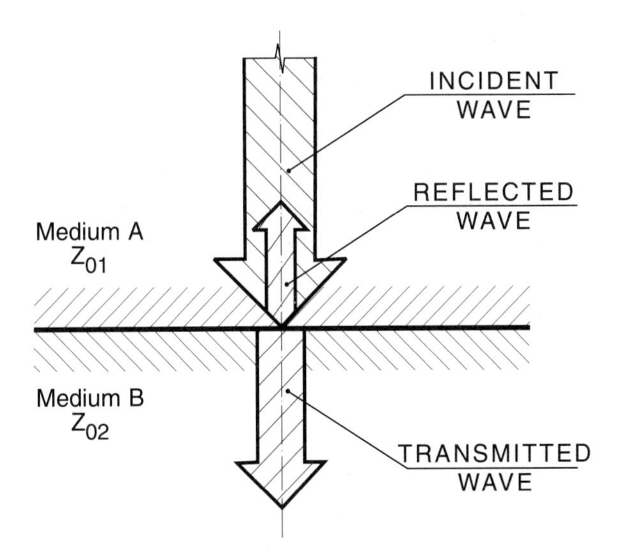

Fig. 2.2 ***Perpendicular incidence of an ultrasound wave on the interface of two media***

The ratio of the intensities of the reflected wave, I_{a2}, to the incident wave, I_{a1}, striking perpendicularly onto the boundary of two media with wave resistances, Z_{01}, and Z_{02}, respectively, is called the reflection coefficient, R_o

$$R_o = \frac{I_{a2}}{I_{a1}} = \left(\frac{Z_{02} - Z_{01}}{Z_{02} + Z_{01}}\right)^2 = \left(\frac{1 - m_z}{1 + m_z}\right)^2 \leq 1 \qquad (2.20)$$

where
$$m_z = \frac{Z_{01}}{Z_{02}} = \frac{\rho_1 c_1}{\rho_2 c_2}.$$

The ratio of the wave intensity, I_{b1}, penetrating into the medium B with a wave resistance, Z_{02}, to the incident wave intensity, I_{a1}, is called the transmission coefficient, D

$$D = \frac{I_{b1}}{I_{a1}} = 1 - R_0 \leq 1 = \frac{4Z_{01}Z_{02}}{\left(Z_{01} + Z_{02}\right)^2} \tag{2.21}$$

In the special case when both the wave resistances are equal ($m_z = 1$), the ultrasonic wave passes undisturbed from one medium into the other. In all other cases, a reflection always occurs.

In practice, it is often necessary to transmit ultrasonic energy from one medium into another with a different wave resistance, ρc. Sometimes it is necessary to separate the media from each other by a layer. In such cases, various limit situations can occur due to the ratio of layer thickness, d, and the wavelength, λ, which affects the transmittability of the ultrasonic wave (Fig. 2.3).

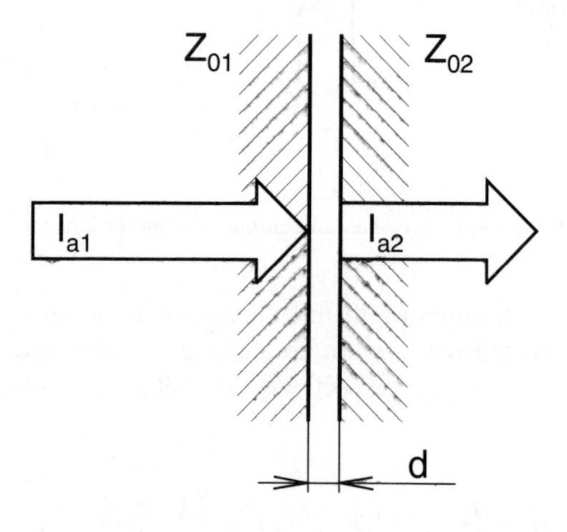

Fig. 2.3 ***Transition of an ultrasound wave through a layer of thickness d***

The maximum reflection ($R_o = 1$) occurs at values of the layer thickness

$$d = (2n - 1)\frac{\lambda}{4} \quad \text{for } n = 1,2,3...$$

If the layer is placed between two different media with wave resistances Z_{01} and Z_{02}, the reflection coefficient is zero ($R_o = 0$) and the transmission is a maximum ($D = 1$) provided that the wave resistance of this coupling and interface layer is

$$Z_0 = \sqrt{Z_{01}Z_{02}} \ .$$

The maximum transmission of the ultrasonic energy occurs at a thickness d

$$d = n\frac{\lambda}{2} \quad \text{for } n = 1,2,3...$$

which corresponds to a layer thickness which is equal to integer multiples of the half-wavelength [2.3].

Maximum transmission of ultrasonic energy also occurs in the case when the layer thickness, d, satisfies the condition $d \ll \lambda$ which means that the transmission of energy from one medium into another is not influenced by the layer thickness ($R_o = 0$, $D = 1$). In practice, a thin membrane corresponds to this case.

These extreme layer thicknesses are important in the design of ultrasonic systems. It is often necessary to produce layers of suitable thickness, especially from the point of view of ultrasonic energy transfer from transmitters into a medium.

2.3 Refraction of ultrasonic waves

In the previous section, a situation with the perpendicular incidence of an ultrasonic wave on the interface of two media was examined. In another situation, a longitudinal wave motion, L, acts on the interface of two media, A and B, at an angle α (Fig. 2.4). A part L_1 of the energy of the wave

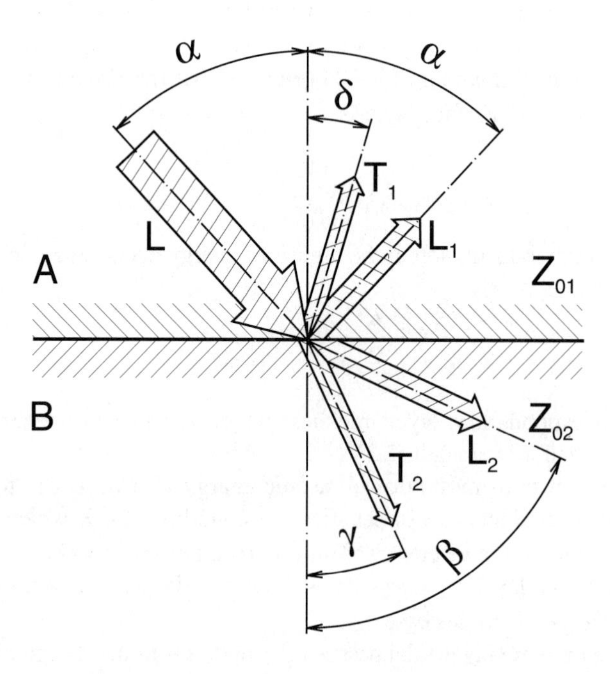

Fig. 2.4 ***Refraction of a longitudinal ultrasound wave at an oblique angle of incidence at the interface of two media, A and B***

motion is reflected back into the medium A, whereby the angle of reflection equals the angle of incidence. Another part of the wave motion, L, penetrates the medium B, and is refracted depending on the mutual wave resistances of the media.

The refraction and reflection of ultrasonic waves are similar to those in geometrical optics. Snell's law of refraction

$$\frac{\sin\alpha}{\sin\beta} = \frac{c_1}{c_2} \tag{2.22}$$

holds true, where c_1 and c_2 are the velocities of propagation of sound in the media A and B, and α and β are the angles of incidence and refraction respectively. The ratio c_1/c_2 has the same meaning for ultrasound as the refractive index does for light. Total internal reflection arises in the cases when

$$\frac{c_2}{c_1}\sin\alpha > 1 .$$

The intensities of the reflected and transmitted waves were given respectively by Rayleigh as

$$I_1 = I\left(\frac{\sqrt{1-\sin^2\alpha} - \frac{\rho_1}{\rho_2}\sqrt{\frac{c_1^2}{c_2^2} - \sin^2\alpha}}{\sqrt{1-\sin^2\alpha} + \frac{\rho_1}{\rho_2}\sqrt{\frac{c_1^2}{c_2^2} - \sin^2\alpha}}\right)^2 \tag{2.23},$$

$$I_2 = I\,\frac{4\frac{\rho_1}{\rho_2}\sqrt{\frac{c_1}{c_2} - \sin^2\alpha}}{\left(\sqrt{1-\sin^2\alpha} + \frac{\rho_1}{\rho_2}\sqrt{\frac{c_1}{c_2} - \sin^2\alpha}\right)^2} \tag{2.24}.$$

From Eq. (2.23), the angle can be found at which total refraction occurs, i.e. when all the energy is transmitted into the second medium. Total refraction occurs at $I_1 = 0$, i.e. when

$$\sqrt{1-\sin^2\alpha} = \frac{\rho_1}{\rho_2}\sqrt{\frac{c_1^2}{c_2^2} - \sin^2\alpha} .$$

Thus, total refraction occurs in the case when the wave acts on a boundary with acoustic impedances $\rho_1 c_1$, and $\rho_2 c_2$, at an angle α, whose value is

$$\alpha = \arcsin\left[\frac{\rho_1^2 c_1^2 - \rho_2^2 c_2^2}{\left(\rho_1^2 - \rho_2^2\right)c_2^2}\right] \qquad (2.25).$$

The reflection coefficient, R_o, is larger for oblique incidence than for perpendicular in the case when the condition $c_1 < c_2$ is satisfied.

Besides refraction and reflection, a partial transformation to other types of wave can occur for oblique incidence at the interface of two media, if for a particular type of wave, some limit angle value is exceeded. This means that apart from the longitudinal wave, L_2, a transverse wave, T_2, or even a surface wave can arise in the second medium. Whilst in perpendicular incidence only longitudinal waves can arise in the second medium, in oblique incidence, with increasing angle α, longitudinal waves L_1, L_2, can also be excited and the proportion of transverse waves T_1, T_2, increases as well, until the longitudinal wave expires [2.4], [2.5].

2.4 Attenuation of ultrasonic wave motion

The propagation of ultrasonic wave motion can be described as elastic deformations of individual particles of a medium. Due to internal friction and thermal conductivity, this deformation is accompanied by losses of oscillation energy, which are converted to thermal energy.

In solids, apart from these losses, reflection, refraction, and scattering of the ultrasonic wave motion arise as well. The type of attenuation considered here is scattering. A typical example for scattering losses are metals which are composed of a large number of randomly oriented grains. In the MHz range of frequencies, a basic part of the attenuation is caused by energy losses through scattering of waves on individual grains. If the average size of the crystalline grains, $\overline{D}$, is about 20 times less than the wavelength, λ,

$$\lambda \geq 20\overline{D}$$

then the attenuation coefficient, α, is a linear function of frequency, f [2.6]. The value α/f is constant. The linear dependence of the attenuation coefficient on frequency ceases to be constant above some limiting frequency resulting from the condition $\lambda \leq 20\,\overline{D}$. Then it can be expressed as

$$\alpha = a_1 f + a_2 f^2 \tag{2.26}$$

where a_1 - absorption losses [s.m^{-1}]
a_2 - scattering losses.[s^2.m^{-1}].

The quantities a_1, a_2 are different for longitudinal and transverse waves. The attenuation of transverse waves is larger than that of longitudinal waves.

If the wavelength, λ, is comparable with the average grain size, $\lambda = \overline{D}$, then so-called stochastic scattering occurs. In this case, the increase in loss is proportional to the grain size, D. As an example, the dependence of the attenuation coefficient, α, on frequency, f, for carbon steel is depicted in Fig. 2.5.

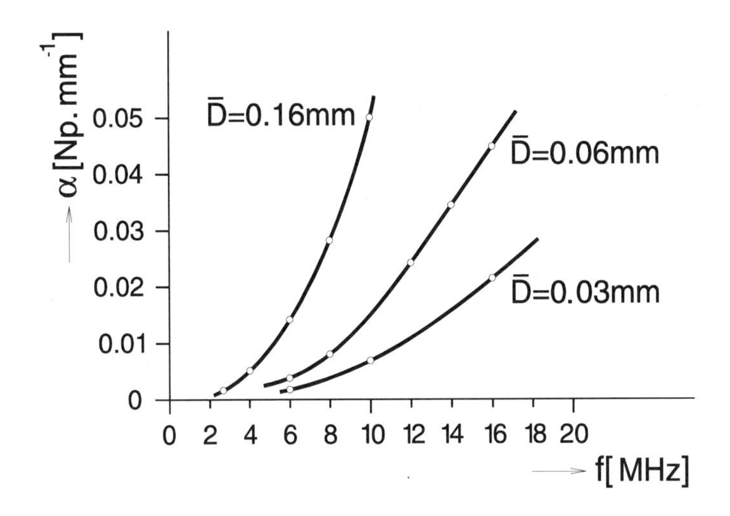

Fig. 2.5 ***Dependence of the attenuation coefficient, α, on frequency, f, for carbon steel***

In glasses and fused quartz, the attenuation coefficient is proportional to frequency. The minimum attenuation occurs in fused quartz whereby its attenuation coefficient for longitudinal waves is higher by some orders of magnitude than that for transverse waves.

The attenuation in plastics is higher by two to three orders of magnitude compared with metals. For instance, the attenuation coefficient for polyethylene at a frequency 1 MHz is about 500 times higher than that for aluminium or fused quartz.

Values of the attenuation coefficient of longitudinal waves, at a frequency of 1 MHz and a temperature 20 °C, for some media are given in Table 2.4.

TABLE 2.4

Attenuation coefficient of longitudinal ultrasonic waves at a frequency 1 MHz and a temperature 20 °C

medium	attenuation coefficient α [$Np.mm^{-1}.MHz^{-1}$]	medium	attenuation coefficient α [$Np.mm^{-1}.MHz^{-1}$]
steel	$5.10^{-4} - 5.10^{-3}$	water	$2.5.10^{-5}$
aluminium	$5.10^{-5} - 2.10^{-3}$	glycerol	6.10^{-3}
magnesium	$3.10^{-5} - 3.10^{-4}$	edible oil	$8.4.10^{-3}$
copper	$10^{-3} - 5.10^{-3}$	olive oil	$1.2.10^{-3}$
fused quartz	$0.6.10^{-4}$	methanol	$3.4.10^{-6}$
lead glass	$3.2.10^{-4}$	ethanol	$5.4.10^{-6}$
window glass	$3.2.10^{-4}$	acetone	7.10^{-6}
organic glass	$2.5.10^{-2}$	benzene	9.10^{-5}
polystyrene	$1.7.10^{-2}$	toluene	20.10^{-5}
polyethylene	$5.2.10^{-2}$	mercury	6.10^{-7}
nylon	3.10^{-2}	air	$1.6.10^{-2}$

In liquids and gases, the attenuation of ultrasound depends on the losses caused by viscosity, α_v, and on the losses caused by the thermal conductivity of liquids, α_T. The attenuation coefficient of ultrasound can be expressed as

$$\alpha = \alpha_v + \alpha_T = \frac{2\pi^2}{\rho c^3}\left[\frac{4}{3}\eta + \lambda_T\left(\frac{1}{c_v} - \frac{1}{c_p}\right)\right] f^2 = af^2 \qquad (2.27)$$

where ρ - density [kg.m^{-3}]
c - velocity of ultrasound [m.s^{-1}]
η - dynamic viscosity [Pa.s]
λ_T - thermal conductivity of medium [W.m^{-1}.K^{-1}]
c_v,c_p - specific heats at constant volume and constant pressure, respectively [J.kg^{-1}.K^{-1}].

It can be seen from (2.27) that the attenuation of ultrasound in liquids and gases is proportional to the square of frequency. The value α/f^2 is constant. For most liquids, the attenuation coefficient, α_T, depending on thermal conductivity, is considerably less than the coefficient α_v, which is caused by the viscosity of the liquid, $\alpha_v \gg \alpha_T$.

On the other hand, in some liquids, such as mercury, the value of α_T is a few times larger than α_v.

In gases, the values of both the attenuation coefficients are approximately equal (at least in order of magnitude), $\alpha_v \approx \alpha_T$.

The significance of Eq. (2.27) is plainly theoretical. The results of measurements show that the calculated values are considerably smaller. This can be explained by a relaxation theory of ultrasound attenuation. This theory is based on the assumption that propagation of wave motion in liquids and gases gives rise to relaxation phenomena and that a certain time is necessary to obtain equilibrium. The losses which arise in this way cause additional attenuation.

The attenuation of ultrasound depends on the medium temperature. With increasing temperature, the velocity of propagation in liquids decreases, and the attenuation increases. The effect is opposite in water.

The attenuation of ultrasound in suspensions depends on the size of the solid particles, and their concentration.

2.5 Focusing of ultrasonic waves

In industry as well as in the laboratory, it is often necessary to concentrate ultrasonic waves, either for the purpose of obtaining high values of signal-to-noise ratio, or high intensities of ultrasonic energy for technological purposes.

For focusing ultrasonic waves, geometrical ultrasonic optics can be used. For this purpose, an ultrasonic wave can be imagined as a direct beam. This assumption is in good agreement with reality in the cases when the boundary surface is much larger than the wavelength of the incident ultrasonic wave. The expressions given for the interaction of the wave with a curved boundary (concave or convex) are valid only under this condition.

In Fig. 2.6 several possibilities of the result of incidence of a planar ultrasonic wave with a cylindrical (concave or convex) interface are illustrated. A perpendicular incident beam moves from medium 1 into medium 2, with a coefficient of reflection R and a coefficient of refraction D. Other beams are refracted by Snell's law into a single point. On the boundary of a spherical shape, all beams intersect in a focal point. In the case of a cylindrical boundary, a focal line is created. A curved boundary can function either as a lens, or as a mirror. According to the value of the refractive index, i.e. whether it is larger or less than one, either a converging or a diverging lens is created, on a convex, or a concave boundary.

For a radius r, of the boundary curvature, the focal length is given by

$$f_0 = \frac{r}{1 - c_2/c_1} \qquad (2.28).$$

This has a positive value if the focus is on the same side as the centre of the curvature.

It is necessary for good ultrasonic lenses that the values of the acoustic wave impedances of both media are about the same. In such a case, the losses arising at the interface are minimal.

2.6 Radiation of ultrasonic energy

A sphere with a radius less than the wavelength of the emitted wave can be considered as a point source, which is the simplest case of radiation of

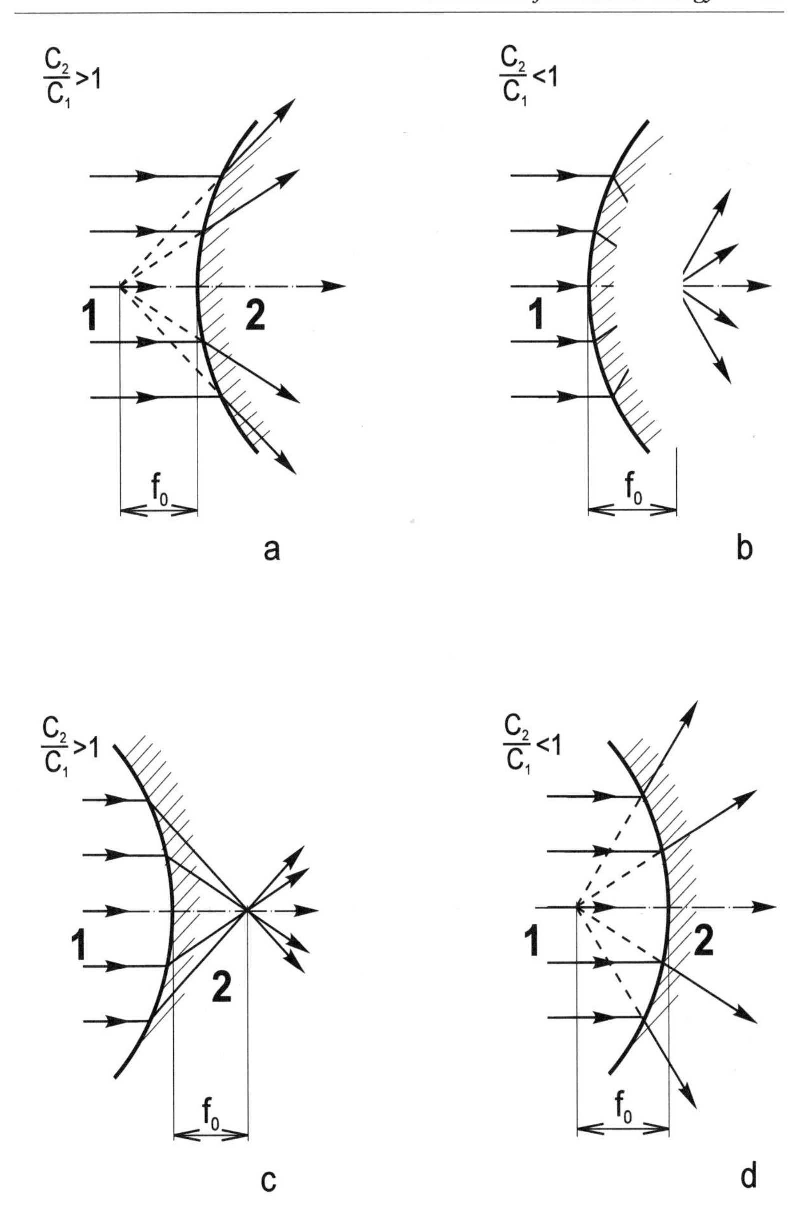

Fig. 2.6 ***Incidence of a planar wave on a curved surface***

an ultrasonic wave. The radiation of ultrasound from a point source is isotropic which gives rise to a spherical wave. A small plate, where all points of its surface oscillate with the same amplitude and phase, is another example of a radiation source. This way of considering the radiation of oscillations is called a plunger method. By Huygens' principle, such a plate can be imagined as being composed of an infinite number of oscillating point sources, each of them radiating a spherical wave (Fig. 2.7).

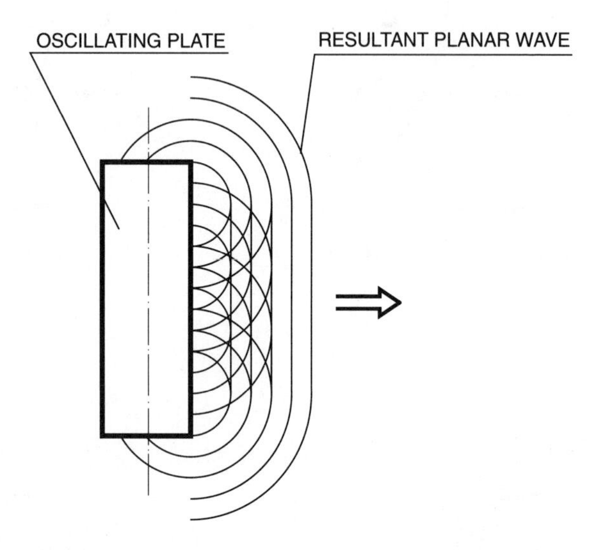

Fig. 2.7 *Radiation field of an oscillating plate according to Huygens' principle*

An elementary area, ds, of the oscillating plate of an ultrasonic transducer excites an acoustic pressure, dp, in the space in front of the transducer face. At a distance r, this pressure is expressed as

$$dp = p_0 ds \frac{e^{-j\omega r/c}}{\lambda r} \tag{2.29}$$

where $p_o = P_o e^{j\omega t}$ - acoustic pressure close to the surface of the oscillating plate

r - distance from the source.

After integration of (2.29) is performed over the whole plate, we obtain for the amplitude of acoustic pressure, P, along the source axis

$$P = 2P_o \sin\left[\frac{\pi}{2}\left(\sqrt{\frac{D^2}{4} + \ell^2} - 1\right)\right] \qquad (2.30)$$

where P_o - amplitude of the original acoustic pressure
D - source diameter
ℓ - distance along the axis.

It follows from (2.30) that the acoustic pressure along the axis exhibits minima and maxima. The maxima are at distances

$$I_{n\,max} = \frac{D^2 - \lambda^2(2n+1)^2}{4\lambda(2n+1)} \quad ; \quad n = 0,1,2... \qquad (2.31)$$

and the minima at distances

$$I_{n\,min} = \frac{D^2 - 4\lambda^2 n}{8n\lambda} \quad ; \quad n = 1,2,3... \qquad (2.32).$$

It can be seen from (2.31) and (2.32) that with increasing n, the maxima and minima get closer to the transducer. The position of the last maximum on the source axis is the most significant one. Its distance from the source is

$$I_o = \frac{D^2 - \lambda^2}{4\lambda} \qquad (2.33).$$

At higher ultrasonic frequencies, when the wavelength is smaller than the diameter of the oscillating plate, $D^2 \gg \lambda^2$, then the distance of the near field, i.e. the position, I_o, of the last maximum of acoustic pressure (for $n = 0$) can be expressed, according to (2.33), as

$$I_o = \frac{D^2}{4\lambda} \qquad (2.34).$$

The existence of maxima and minima of the acoustic pressure, which exist not only on the axis but throughout the whole near field, can be explained by the interference of ultrasonic oscillations arriving at a certain point from different point sources with different phases. At any point, the acoustic pressure results from amplitudes and phases of individual partial waves. The amplitude and phase situation is rather complicated within the near field. A characteristic feature of the near field is that the shape of its cross-section is approximately similar to the shape of the source. This means that there is practically no divergence of the beam.

If the distance is larger than the position of the last acoustic pressure maximum on the source axis, then a so called far field begins (the Fraunhofer region). In this region, the acoustic pressure decreases uniformly with the distance from the source. In contrast to the near field, the cross-section of an ultrasonic beam in the far field does not remain constant, but the beam becomes divergent.

The far field begins theoretically at the end of the near field. In real situations, there is always a transient zone between them.

2.7 References

2.1 *Blitz J.:* Fundamentals of Ultrasonics, Butterworth, London, 1963
2.2 *Taraba O.:* Vybrané stati z fyzikální akustiky, 1. Ultrazvuk, ČVUT, Praha, 1972 (in Czech)*
2.3 *Truel R., Elbaum Ch.:* Ultrasonic methods in Solid State Physics, Academic Press, New York, 1969
2.4 *Goldman R.G.:* Ultrasonic Technology, Reinhold Publ. Co., New York, 1962
2.5 *Frederick J.R.:* Ultrasonic Engineering, Reinhold Publ. Co., New York, 1965
2.6 *Mason P.W., McSkimin H.J.:* Attenuation and Scattering of High Frequency Sound Waves in Metals and Glasses, J. Acoust. Soc. Amer. 19, 1947, 466

* Selected parts of physical acoustics, 1. Ultrasound

3 Ultrasonic transducers

3.1 Capacitive transducers

3.1.1 Physical principle of operation

Transducers can be used as transmitters (generators) or as receivers (detectors) of ultrasonic waves. They exploit a mutual interaction between the stationary (fixed) electrode, and the moving electrode (membrane) [3.1], [3.2].

In transmitters, the effect of electrostatic forces on a moving electrode is exploited. A mechanical force causes deflection of the electrode and thus the pressure changes in its vicinity as well. A basic deflection of the membrane is achieved by connection of a polarisation (bias) voltage, V_p, which results in attractive electrostatic forces. An ac voltage term v is superimposed on this basic bias voltage.

In receivers, an incoming sound wave causes a change in the capacitance between the electrodes. In the so called constant charge mode, V_p is connected through a high resistance. In this way, a change in capacitance is transformed into an electrical voltage signal.

The transducers can operate in reciprocal fashion, and they can be fabricated in various shapes to increase the sensitive surface area (within technological constraints).

The basic principles and design of a capacitor transducer are well known, and theoretical solutions are described in detail in the literature.

In the following pages, some results of a theoretical nature [3.3] are briefly introduced.

An elementary force, dF, acts on an area element, dS

$$dF = \frac{\varepsilon_0 dS \left(V_p + v\right)^2}{2\left(d - \eta\right)^2} \tag{3.1}$$

where d - rest distance between the electrodes

η - deflection of the membrane.
v - applied voltage.

This yields the (acoustic) pressure, p:

$$p = \frac{dF}{dS} = \frac{e_0 \left(V_p + v\right)^2}{2\left(d - \eta\right)^2} \qquad (3.2).$$

The total differential of p for variables v and η is

$$dp = \frac{\partial p}{\partial v} dv + \frac{\partial p}{\partial \eta} d\eta$$

and using 3.2, we obtain

$$dp = \frac{\varepsilon_0 \left(V_p + v\right)}{\left(d - \eta\right)^2} dv + \frac{\varepsilon_0 \left(V_p + v\right)^2}{\left(d - \eta\right)^3} d\eta \ .$$

Supposing v « V_p , and η « d, which in practice is satisfied and neglecting the term for the pressure due to the bias voltage alone, one can write for the pressure, p

$$p = \frac{\varepsilon_0 V_p}{d^2} v + \frac{\varepsilon_0 V_p^2}{d^3} \eta \qquad (3.3).$$

Introducing a coefficient k_β

$$k_\beta = \frac{Q_0}{Y_0} = \frac{C_0 V_p}{Y_0} = \frac{\varepsilon_0 S V_p}{d Y_0} = \frac{\varepsilon_0 V_p}{d^2} \qquad [C.m^{-3}]$$

where Q_o - charge on the transducer capacitance ($Q_o = C_o V_p$)
C_o - static capacitance
Y_o - volume between the electrodes ($Y_o = Sd$)
S - total membrane area

whereby $$\frac{\varepsilon_0 V_p^2}{d^3} = \frac{\varepsilon_0^2 V_p^2 d}{d^4 \varepsilon_0} = k_\beta^2 \frac{d}{\varepsilon_0}$$

yields

$$p = k_\beta v + k_\beta^2 \frac{d}{\varepsilon_0} \eta \qquad (3.4).$$

Now, the equation 3.4 can be multiplied by dS and integrated over S

$$pS = k_\beta S v + k_\beta^2 \frac{d}{\varepsilon_0} \iint_S \eta \, dS \quad .$$

Supposing the harmonic motion, $\mathbf{w_0} = j\omega Y$,
where $\mathbf{w_0}$ - volume acoustic velocity; $[m^3.s^{-1}]$ is equal
$|\mathbf{w_0}| = w_0 = \omega Y = \omega S \eta$,

then integral $$\iint_S \eta \, dS = Y = \frac{\mathbf{w_0}}{j\omega} \quad ,$$

we obtain

$$p = k_\beta v + \frac{k_\beta^2}{j\omega C_0} \mathbf{w_0} , \qquad \text{where } C_0 = \frac{\varepsilon_0 S}{d}$$

and further introducing: $$\frac{C_0}{k_\beta^2} = c_{n\beta} \qquad [m^5.N^{-1}]$$

where C_o - static capacitance
$c_{n\beta}$ - negative acoustic compliance,
yields

$$p = k_\beta v + \frac{1}{j\omega c_{n\beta}} \mathbf{w_o} \qquad (3.5).$$

The electric parameters follow from the relationship between charge and voltage. An elementary charge, dq, on an area dS, can be expressed as

$$dq = (V_p + v)\, dC,$$

where
$$dC = \frac{\varepsilon_0 dS}{d - \eta}$$

then
$$q = \varepsilon_0 (V_p + v) \iint_S \frac{dS}{d - \eta}.$$

It is reasonable to assume $\eta \ll d$. Then, d can be taken in front of the integral and the rest,

$$\frac{1}{1 - \eta/d}\, dS$$

can be substituted with sufficient precision by the expression $(1 + \eta/d)dS$. Then,

$$q = \frac{\varepsilon_0 \left(V_p + v\right)}{d} \iint_S \left(1 + \frac{\eta}{d}\right) dS \qquad (3.6).$$

After forming the derivative $\frac{\partial q}{\partial t} = i$,

$$\frac{\partial q}{\partial t} = \frac{\partial v}{\partial t} \frac{\varepsilon_0}{d} \iint_S (1 + \eta/d)\, dS + \frac{\varepsilon_0 \left(V_p + v\right)}{d^2} \iint_S \frac{\partial \eta}{\partial t}\, dS ,$$

where for harmonic signals, $\frac{\partial v}{\partial t} = j\omega v$.

The first integral, for values $\eta/d \ll 1$ approximately equals S, and the second integral

$$\iint_S \frac{\partial \eta}{\partial t}\, dS = \iint_S \bar{v} dS = w_0$$

where $\bar{v}$ is the acoustic velocity. Then the final form of the equation for i is

$$i = j\omega C_o v + k_\beta \mathbf{w}_o \qquad (3.7).$$

The equations (3.5) and (3.7) are a fundamental starting point when drawing equivalent circuit diagrams and calculating some of the transducer parameters, e.g. resonance frequency, slope discontinuities in the frequency characteristics or microphone sensitivity. This theory and the expressions derived are valid for a single-action transducer type, where a membrane with well-known and stable qualities (essentially a metallic one) is fastened on the perimeter. Actual designs of ultrasound transducers vary from this ideal, and the foregoing solutions can be used only approximately.

3.1.2 The design and qualities of an ultrasonic transducer

For higher frequencies, the mass of the moving parts must be reduced considerably in order to obtain a higher resonance frequency and higher frequencies of the slope discontinuities in the frequency characteristics. Different materials must be used and the overall dimensions of the oscillating parts should be reduced.

Plastic foils are used as membranes, e.g. styroflex, mylar (polyethyleneterephtalate), and others, with thicknesses of 3 – 40 μm. The foils are metalized on one side by a very thin layer of gold, copper, or aluminium. Exceptionally, metallic membranes, created by special techniques, are used too. For this purpose, fine metals are commonly used, e.g. nickel and titanium, in thicknesses of units of micrometres.

Furthermore special plastic foils of the electret type can be used, that are polarized internally. These transducers do not need a bias voltage.

Both the membrane and the complete transducer can also be fabricated by the use of integrated circuit technology in silicon. For example, a membrane created of silicon nitride may have dimensions of 0.8 x 0.8 mm, and a thickness of 150 nm, with a 100 nm aluminium layer.

The effort in all the designs is to keep the working gap as small as possible over the whole transducer membrane. This can hardly be achieved by tensioning the membrane or by similar means. A more suitable way is to divide the sensitive area into a larger number of smaller elementary transducers. This can be accomplished by a system of regularly arranged grooves, sinks and elevations, on the live electrode. Some of the possible configurations are depicted in Fig. 3.1a,b,d. Apart from this, the surface can be roughened irregularly, e.g. by sand blasting, (Fig. 3.1.c).

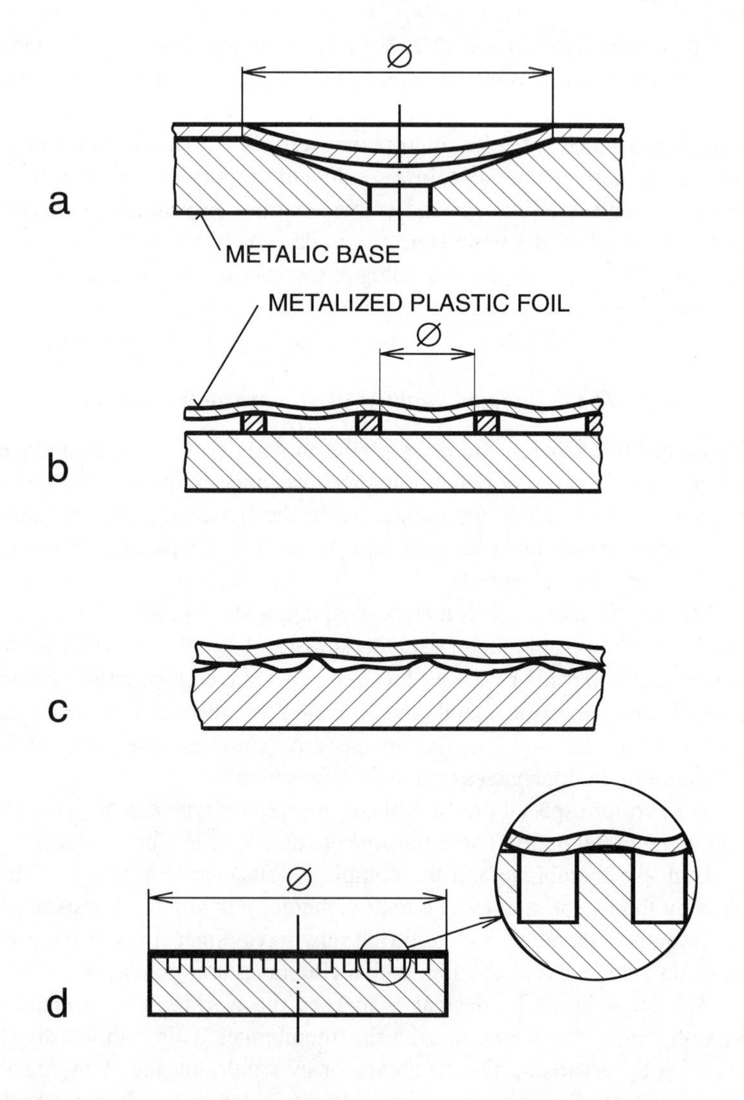

Fig. 3.1 ***Arrangement of the active parts (electrodes) of capacitive transducers***

3.1.3 Shape of the sensitive area and the scope of utilization

The shape of the sensitive area can be:

a) flat - large surface

b) a cylinder or a part of a cylindrical surface

c) spherical

d) a point or quasi-point.

Systems a) are used in distance measurement, or in straight line coordinate measurements, in remote control and in the transfer of limited amounts of information. They can be created by repetition of any element from Fig. 3.1. Their advantage is the ability to direct waves, and consequently they have a large operating range, up to tens of meters.

Systems b) are used in distance measurement, or coordinate measurement, in a straight line as well as in a plane. They can be created by the repetition of any element from Fig. 3.1. on a cylindrical surface. Their radiation pattern is omnidirectional (in one plane) which leads to a dissipation of the transmitter energy and a reduction in the sensitivity of the receivers. This results in a shortening of the operating range.

Systems c) and *d)* can be used in measurements in a straight line, in a plane, and in space. The radiation pattern is omnidirectional in space; thus, their operating range is still shorter.

The creation of a point transducer based on a capacitive system is unrealistic. Theoretically, it is possible to create a spherical source with finite dimensions which could be corrected by calculation. In practice, the creation of a spherical surface represents a problem that can hardly be mastered by technology.

Mutual combinations of these systems are also possible [3.4].

3.1.3.1 The principle of a quasi-point transducer – a microphone

For some purposes, a sensitivity in the spatial angle of less than 2π is sufficient, for instance for the detection of an ultrasonic wave in picking-up spatial coordinates. For this task, a hemispherical shape, or a so called quasi-point transducer is satisfactory. This is most often used in a so called receiver-microphone mode.

The considerations start with the relationship of the size (diameter ϕ) of the sensitive area to the measured distance, l, and from the error created in this way. The principle is illustrated in Fig. 3.2.a.

Supposing the whole area, ϕ, is sensitive, then on misalignment of say, point B, the shortest path for sound is l'. Since $l' < l$, the arrival of the

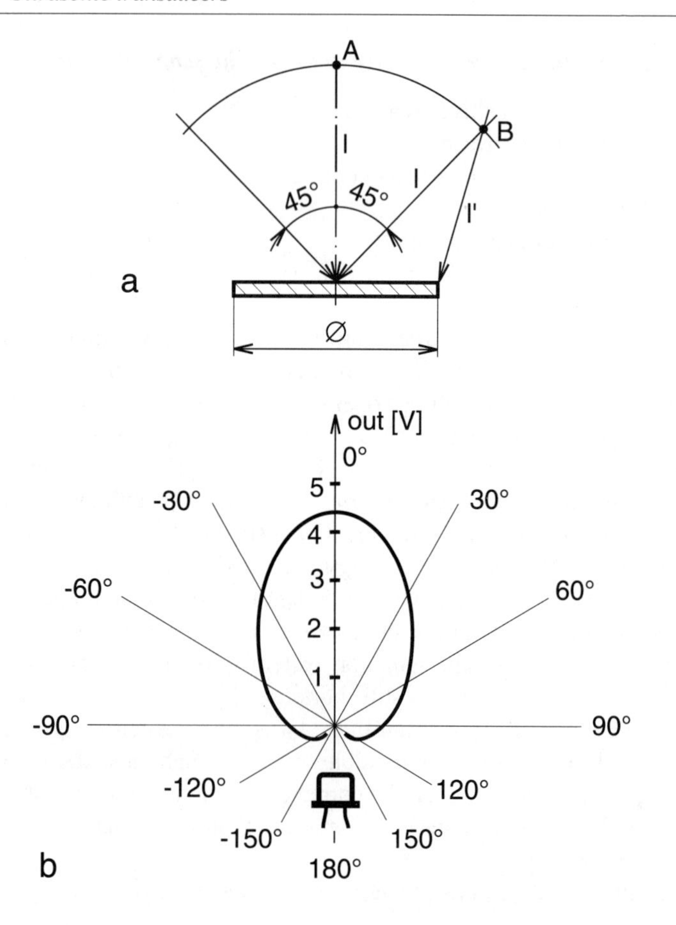

Fig. 3.2 ***Illustration of the origin of an angular error (a), and a directional characteristic (b), of a quasi-point shaped capacitor microphone***

sound wave will be detected by the microphone earlier than from the point A. For a microphone diameter $\phi = 2$ mm, an angular deviation of 45° and a distance of 50 mm, the absolute error is about 0.7 mm (1.4 %). For larger distances, the value of the relative error decreases while the absolute error stays almost unchanged.

The sensitive area of a microphone can be lessened and, by this means, the error decreases. However, the sensitivity, and consequently the range also, decreases at the same time. A size of the sensitive area of

$\phi = 2$ mm, in connection with a spark source of sound, is sufficient for a range of 5 - 10 m, as determined by experiment.

These receivers are considered to be microphones of zero order; they are sensitive to the acoustic pressure, p. However, their directional radiation sensitivity pattern is not purely spherical (or circular in a plane), because they distort the acoustical field. The characteristics resemble a cardioid-like shape, with increased sensitivity in the forward direction (Fig. 3.2.b).

An example of one particular design of a quasi-point microphone with given characteristics is shown in Fig. 3.3.

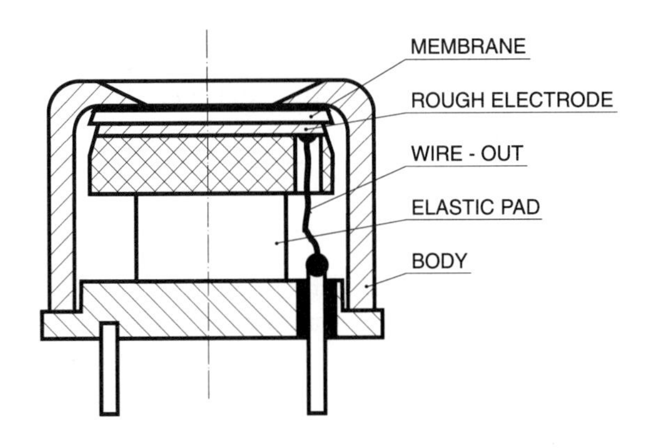

Fig. 3.3 ***Simplified design for the quasi-point capacitive transducer***

3.1.3.2 The design and associated analytical solution for transducers

The design and the associated analytical solution for transducers for ultrasound frequencies is not easy. The membrane is the critical point. Plastic membranes are sensitive to temperature and humidity changes so that their dimensions and their mechanical properties change over a wide range. Consequently, the parameters and constants necessary for calculations change as well. Thus the problem of appropriate tensioning of the membrane results. It is recommended that a loosely laid, mechanically non-tensioned membrane is used, held only by electrostatic forces resulting from the polarisation (bias) voltage. The value of the polarization voltage depends on the required deflection of the membrane. For the upper value, the limiting factor is the breakdown strength of the membrane and its thickness, and the voltage is about 300 V, as a rule. With increasing polarization voltage, the sensitivity of the microphones increases. When using electrets, the need for a polarization voltage disappears.

The live electrode represents another critical point. Here, the problem lies in maintaining tolerances, and the reproducibility of the technological process for machining its surface.

The upper frequency limit depends particularly on the membrane. The frequency range and the sensitivity depend mostly on the live electrode. For any design, its purpose and the scope of the application determine this, as well as the type of excitation, i.e. whether the transducer works in a continuous or in a pulsed mode. It is partly possible to use some formulae valid for the audible frequency range; however, these approximate solutions must be corrected and checked by practical experiments.

3.1.4 Excitation and evaluation circuits

The basic modes of transducer operation are (a) continuous and (b) pulsed. These relate to both transmitters and receivers. In continuous mode, the most suitable excitation is via a periodic voltage. The choice of frequency depends on the requirements placed on the method. Unless the demands are specified, it can be a resonant frequency of the transducer. A non-harmonic shape (e.g. rectangular which is usually easier to generate), can also be used. The base harmonic frequency must equal the required excitation frequency. In both cases the application of a polarization voltage, on which the excitation voltage term is superimposed, is also required.

In pulse mode an excitation by a radio pulse is used, with a suitable, prescribed carrier frequency (sine, rectangular), that again is superimposed on the polarization dc voltage term. In pulse mode, also, a single broad excitation pulse can be used. In this case, the transducer would exhibit a few damped oscillations at its resonance frequency.

The amplitude of the excitation signal (at the transducer) is usually 10 - 300 V_{pp}, and the polarization voltage is usually 100 - 300 V.

The principal circuit diagram of the excitation circuitry, which is also applicable for the reciprocal operation of reception, is shown in Fig. 3.4.

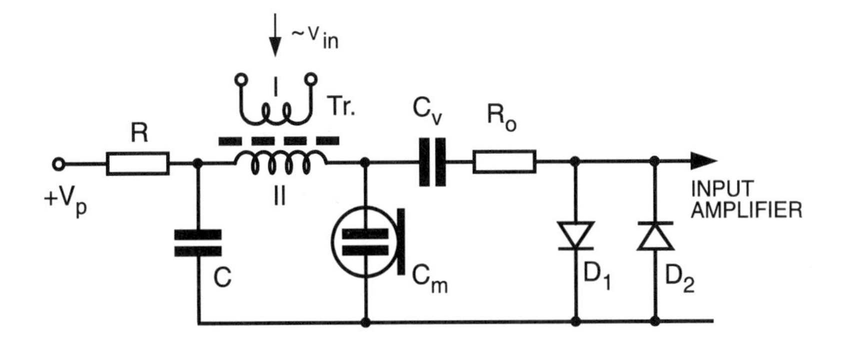

Fig. 3.4 ***Circuit diagram of an excitatory circuit of a capacitive transducer***

In transmission, the ac voltage term $\sim v_{in}$, connected to Tr, is superimposed on the dc polarization voltage, V_p. The capacitor C serves as a filter and it closes the circuit of the ac voltage term. Without the polarization dc term, the transducer would tend to transmit at doubled frequency and with high distortion. When using electrets, the polarization term is not necessary. In transmission, the amplifier input is protected by diodes. The amplifier must be sufficiently resistant to saturation, or it must have a short recovery time. In reception, the signal from C_M passes to the amplifier input, and because of its small amplitude, it is not influenced by the diodes. Apart from the common demands on the microphone amplifier (amplification of 10^4 - 10^5, high input impedance of the order of MΩ, frequency range of hundreds of kHz), it also should be immune to interference. This is an important operational limiting parameter, especially in the case of a spark source. For high input impedance and real (small) measuring ranges, this represents something of a problem. It is recommendable to use discrete elements rather than integrated circuits, and the input impedance should be chosen to be lower, even at the cost of a loss of sensitivity. The amplifier should be placed as close as possible to the transducer. The capacitor C_v separates the amplifier input from the polarization voltage, and it must be matched to this voltage.

3.2 Piezoelectric transducers

Piezoelectric transducers have a significant place in the field of ultrasonic measuring techniques. In 1880, the brothers Pierre and Jacques

Curie found that some substances can release a surface electric charge, if subjected to a mechanical strain. This effect was called *pressure electricity* - piezoelectricity, (in Greek, *piezo* = to press).

Materials with natural piezoelectric properties, e.g. quartz, tourmaline and Seignette salt are abundant in nature, mostly in the crystalline form. Their ions, which form a crystalline lattice, are ordered in such a way that if a crystal is deformed, the negative and the positive ions shift, so that each element of the crystal acquires an electric dipole moment, and a charge appears on the whole crystal. This phenomenon is known as a direct piezoelectric effect. The resulting electric voltage on the crystal is proportional to the strain, and it depends on the strain direction, because the piezoelectric crystals are anisotropic. The reverse of the direct piezoelectric effect is the indirect piezoelectric effect, when a mechanical deformation arises in a crystal in an electric field (again, directly proportional) [3.5], [3.6].

In practice, crystalline ferroelectrics, like barium titanate and zirconate ceramics, which exhibit strong electrostrictive properties, are widely exploited. The behaviour of ferroelectrics in an electric field is similar to the behaviour of ferromagnetics in a magnetic field. Ferroelectric ceramics have a strong remanent polarization which can be induced and maintained artificially. After polarization their electroacoustical properties resemble the piezoelectric ones, although electrostriction is rather analogous to magnetostriction.

The relationship between the mechanical and electrical properties of piezoelectric materials is described by the piezoelectric constants. A deformation, ε, connected with an oscillation of a quartz piezoelectric transducer is linearly proportioned to the intensity of the exciting electric field, E:

$$\varepsilon = d_{33}E \tag{3.8}$$

The piezoelectric constant, d_{33}, of quartz is very small, as is the electromechanical coupling coefficient, k. On the other hand, a quartz transducer exhibits high mechanical strength, chemical resistivity and a low temperature dependence of its piezoelectric properties and of its resonance frequency.

In contrast, the deformation ε of crystalline ceramic materials, excited by an electric field E, is

$$\varepsilon = d_{33}E^2 \qquad (3.9)$$

where d_{33} is the piezoelectric constant of the particular material.

Contrary to the quartz transducer, the deformation of a crystalline material transducer increases with the square of the electric field intensity, E. For this reason it is possible to use a lower supply voltage than for quartz materials where ε increases linearly with E [3.7].

The electromechanical coupling coefficient, k, is the next important constant. It is a measure of the transformation of electrical energy into mechanical energy and vice versa. As a rule, it is defined by its square, k^2, which is equal to the ratio of the mechanical energy produced, E_m, to the total of the electrical energy supplied, E_e, at considerably higher frequencies than the resonance frequency, $k^2 = E_m/E_e$. It is expressed as an absolute number, or in a percentage form. However, the electromechanical coupling coefficient does not reveal anything about the efficiency of energy conversion in an oscillatory mode.

In ultrasound transmitters, a low value for the electromechanical coupling coefficient indicates that a higher value of the exciting electric voltage should be applied to the transducer in order to radiate the same intensity of ultrasonic energy. Experience has shown that for a quartz transducer with a coefficient k = 0.1, a supply voltage about 100 times higher must be applied compared with a lead zirconate transducer. In receiver mode, the transducer with a higher coefficient k yields a higher value of electric voltage, and is therefore in both cases (transmitter and receiver) considered more sensitive.

The electromechanical coupling coefficient, k, is defined as

$$k^2 = d_{ik}h_{ik} \qquad (3.10)$$

where d_{ik} is the piezoelectric coefficient which denotes the indirect piezoelectric effect. The constant, h_{ik}, is a piezoelectric deformation constant which characterizes the direct piezoelectric effect. Another constant which describes the sensitivity of a transducer to an external pressure, is the piezoelectric pressure constant, g_{ik}, which is extremely important in the choice of a suitable piezoceramic for receiving probes for measuring purposes. In practice, when employing the same transducer for both reception and transmission of ultrasonic energy, it is necessary for both the coefficients d_{ik} and g_{ik} to have high values, so that an efficient conversion from electric energy into mechanical energy, and vice versa, is achieved. This confirms the fact that the electromechanical coupling coefficient, k, depends directly on these constants and Young's modulus, E_p, through

$$k^2 = g_{ik} \cdot d_{ik} \cdot E_p \quad (3.11).$$

3.2.1 Piezoelectric materials

With respect to the advantageous properties of piezoceramics in practical use, it is worth describing briefly this kind of material. The piezoceramic is produced by pressing a fine powder, and heating it in a tunnel furnace. In this way, various shapes of transducers can be produced: discs, rings, rods, bars, curved shapes, etc. Afterwards, the burnt ceramic is ground and equipped with baked-on electrodes (Ag, Ni, etc.). In this way, a material with electrostrictive properties is obtained. If this material is further polarized, it acquires properties similar to those of piezoelectric materials. The polarization is usually carried out in an oil bath at a temperature slightly above the Curie temperature. The electrodes are connected to a high voltage, reaching a field of about 3×10^6 V.m^{-1} over one hour.

After this, the element is slowly cooled, and the voltage supply remains connected until the temperature is lowered down to 30 °C.

In most ultrasonic measuring techniques, zirconates are usually used. They have higher operating temperatures, up to 200 °C, and their Curie temperature is about 320 °C. In contrast, the Curie temperature of the barium titanates is only 140 °C. Some polymers also exhibit piezoelectric properties. In Table 3.1, the parameters of some significant piezomaterials are introduced.

TABLE 3.1

Constants of some selected piezoelectric materials

material	density $\rho.10^{-3}$ [kg.m^{-3}]	velocity of propagation of longitudinal waves C_L[m.s^{-1}]	acoustic wave resistance $\rho.C_L.10^{-6}$ [Pa.s.m^{-1}]	Curie temperature Θ_C [°C]	piezoelectric coefficient d.10^{12} [m.V^{-1}]	electromechanical coupling coefficient k
quartz SiO_2	2.65	5 760	15.3	576	$d_{11} = 2.3$	$k_{11} = 0.1$
zirconate piezoceramics PZT-5A	7.75	3 880	30	365	$d_{33} = 374$	$k_{33} = 0.75$
lead-niobate piezoceramics $PbNb_2O_6$	5.8	2 800	16	550	$d_{33} = 80$	$k_{33} > 0.4$
piezopolymer PVDF	1.78	2 260	4.6		$d_{31} = 25$	$k_{33} = 0.2$
piezopolymer P (VDF-TrFE)	1.88	2 400	4.51		$d_{31} = 12.5$	$k_{31} = 0.12$

3.2.2 An equivalent circuit diagram of a piezoelectric transducer

Piezoelectric transducers in the close vicinity of a resonance behave like an electric LCR series oscillating circuit, with a static capacitance C_s for the transducer connected in a parallel manner (Fig. 3.5). The static capacitance, C_s, represents a capacitor whose capacitance depends on the transducer thickness, permitivity, type of piezoelectric material, and the area of the electrodes. The series LCR circuit represents the moving part of the transducer.

By analysing the equivalent circuit diagram of the transducer as well as by measuring the frequency dependence of the impedance, two significant resonance frequencies can be identified in the resonance region. These correspond to the serial and the parallel resonances.

The serial resonance is characterized by a maximum admittance, Y, and a minimum impedance, Z (Fig. 3.6). It is expressed as

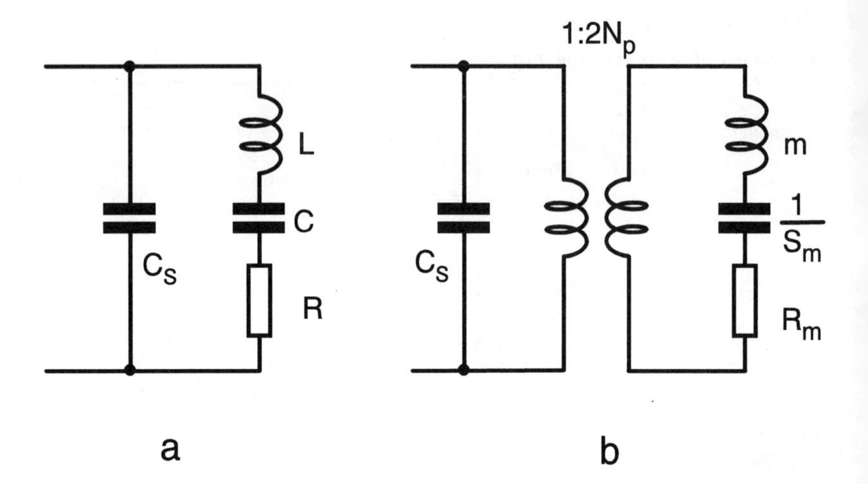

Fig. 3.5 ***Equivalent electrical (a) and electromechanical (b) circuits of a piezoelectric transducer***

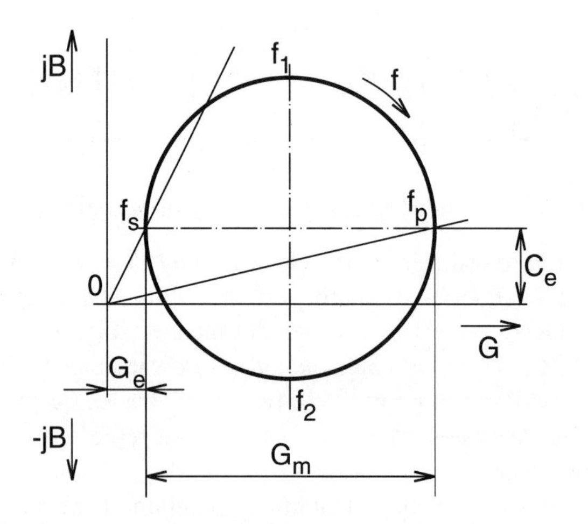

Fig. 3.6 ***The admittance diagram of a piezoelectric transducer (G – the real term of conductivity, jB – the imaginary term of conductivity, f_S – serial (resonance) frequency, f_P – parallel (antiresonance) frequency***

$$f_s = \frac{1}{2\pi\sqrt{LC}} \qquad (3.12).$$

The parallel (or antiresonant) resonance is characterised by a minimum admittance and a maximum impedance, Z, and is given by

$$f_p = \frac{1}{2\pi} \cdot \sqrt{\frac{C + C_s}{LCC_s}} \qquad (3.13).$$

In the frequency band close to resonance, the impedance behaviour of a piezoelectric transducer can be usefully represented by circular impedance (or admittance) diagrams (Fig. 3.6).

Loading of the transducer output by the surrounding medium is represented on the circular impedance diagram by a change of the circumference diameter and by a shift of the diagram curve. In this case, the parallel impedance decreases and the serial one increases. If the transducer is fully loaded, only the static capacitance, C_s, shows up.

3.2.2.1 The quality coefficient

In ultrasonic measuring techniques, piezoelectric transducers are often operated in pulsed mode. This requires sufficient bandwidth from the transducer. Its measure is a quality coefficient, Q, which can be defined by the expression

$$Q \cong \frac{f_r}{f_1 - f_2} \cdot \frac{f_r}{B} \qquad (3.14),$$

where f_r - transducer resonance frequency
f_1, f_2 - frequencies characterized by a decrease of amplitude by 3 dB, relative to the resonance frequency
B - bandwidth.

The quality coefficient, Q_m, of the transducer falls with increasing mechanical load, and at resonance it may be expressed as

$$Q_m = \frac{\omega_r L}{R} = \frac{\pi}{2} \cdot \frac{\rho c}{(\rho_1 c_1 + \rho_2 c_2)} \qquad (3.15),$$

where R_e is an equivalent resistance, and ρc, $\rho_1 c_1$, $\rho_2 c_2$ are the acoustic wave resistances of the transducer, of the damping surrounding medium and of the load, respectively.

The foregoing expression starts from a simplified equivalent circuit diagram, and therefore it is usable only for an unloaded transducer, whose quality coefficient is high. In pulsed mode, transducers with a quality coefficient $Q_m > 10$ are required, and expression (3.15) has, in this case, only very approximate validity.

In order to obtain high sensitivity, the circuits are designed in such a way that the static capacitance, C_o, of the transducer is considered a part of the tuned (oscillator) circuit. The electrical quality coefficient of the piezoelectric transducer circuit is then given by the expression

$$Q_e = \frac{R_e}{\omega_r L_p} \tag{3.16}$$

which naturally leads to an increase of the quality coefficient. In many cases, this is not the most suitable solution for pulsed operation. Therefore if very short pulse operation is necessary, it is better not to connect the piezoelectric transducer into the tuned circuit. Then the bandwidth is determined only by the quality coefficient, Q_m.

3.2.3 Polymeric piezoelectric materials

Recently, in the field of ultrasonic measuring techniques, polymeric piezoelectric materials have gained a significant place due to some of their interesting properties. Although generally they could be classified as being among the classical piezomaterials, a subchapter is devoted to them by virtue of their importance.

After the discovery of a strong piezoelectric effect in high-molecular polyvinylidentfluoride (PVDF) by Kuwaui in 1969, this material has found its application in many branches of technology, especially in the production of electroacoustic transducers. These piezopolymeric materials are available in the form of foils. They acquire their piezoelectric properties after polarization. A common polarization method consists of the following operations [3.8]:

1) a single-sided stretching of the foil, 3 to 4 times;
2) a double-sided metallization of the foil;

3) application of an electric field (E = 108 V.m-1) at an elevated temperature (T = 80 to 100 °C);
4) cooling the foil over a period of about 30 minutes while the field remains switched on.

In Table 3.2. the properties of piezoceramics and the piezopolymer are compared. Each parameter is estimated qualitatively as high or low, and denoted by a sign + (advantage), or - (drawback). In Table 3.1, actual values for some piezopolymeric types are given.

TABLE 3.2

Comparison of properties of piezoceramics and piezopolymer

parameter	*sign*	*ceramics*	*polymer*
acoustic wave resistance	ρ_{oL}	high (-)	low (+)
coupling coefficient	k_{33}	high (+)	low (-)
sensitivity to parasitic modes		high (-)	low (+)
Curie temperature	Θ_c	high (+)	low (-)

One advantage of the piezopolymeric foils is that they are amenable to shaping. In addition, their pulse response is much cleaner than that of PZT (zirconate piezoceramics), where the pulse shape can be distorted by unwanted oscillation modes [3.9].

At present, the properties of piezopolymeric materials are being constantly improved, which fuels the expectation of reaching a quality as high as in the classical piezoelectric materials.

3.3 Magnetostrictive transducers

If a ferromagnetic material changes its dimensions due to magnetization, this effect is called magnetostriction. There is a group of ferromagnetic materials which are capable of greater or lesser degrees of magnetostriction. A classical magnetostrictive material is nickel. Its alloys have some significant magnetostrictive properties too. The most important magnetostrictive materials are metals like nickel, cobalt, permalloy (50 % Fe + 50 % Ni), permendur (49 % Fe + 49 % Co + 2 % V) and ferrites.

It is necessary for magnetostrictive materials to have a high electromechanical coupling coefficient, k, and minimal eddy current losses, also at high frequencies. Therefore the magnetization hysteresis loop should be narrow, and the electrical resistance of the material as high as possible.

Like electrostriction, magnetostriction is anisotropic too. It is manifested as a one-dimensional elongation. As a result, at a frequency f of the exciting ac magnetic field, a mechanical deformation appears at double the frequency, 2f (Fig. 3.7). If a stationary field with an intensity H_o

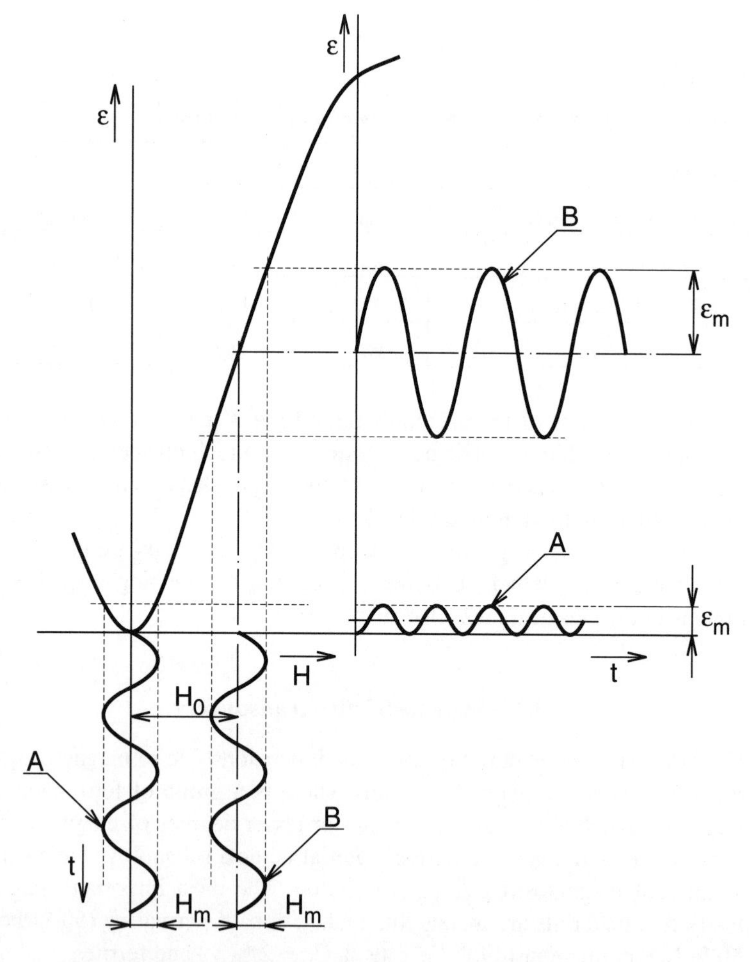

Fig. 3.7 ***The dependence of the linear elongation of a magnetostrictive transducer on the magnetic field intensity during an excitation, without polarization (A), and with polarization (B)***

is introduced, then the total intensity of the ac excitation field is

$$H = H_o + H_m \,.\, \sin\omega t \,.$$

In the linear working region, the time dependence of the deformation is

$$\varepsilon = \beta\mu_i \, H \,.\, \sin\omega t$$

where $\beta\mu_i$ is a measure of the magnetostrictive deformation.

The amplitude of the deformation ε is much higher in a pre-magnetized, i.e. a polarized, transducer than in a transducer without pre-magnetization, but the frequencies of the electrical oscillations are the same.

In metallic transducers the polarization is performed by a dc current, whereby the source of the exciting ac current and the source of the polarization current are separated by a capacitor and a reactance coil. It is simpler to polarize using a permanent magnet whose flux is contained within the core of the transducer (Fig. 3.8).

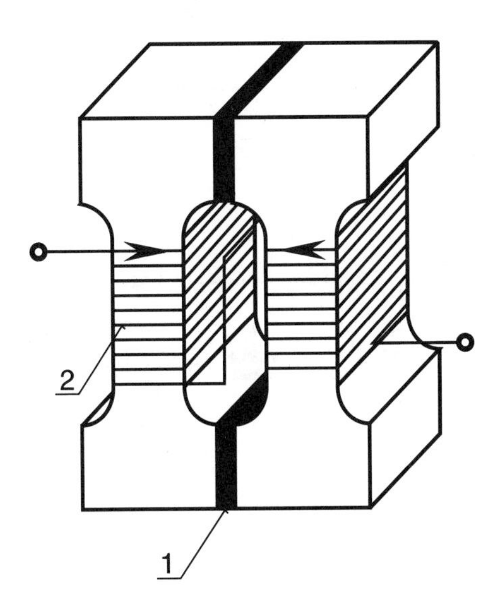

Fig. 3.8 ***A magnetostrictive transducer with a polarization produced by a permanent magnet (1 – permanent magnet, 2 – excitatory winding)***

Magnetostrictive transducers produced from metallic materials are fabricated from sheets of thickness 0.05 - 0.2 mm, in order to minimize eddy currents. For ultrasonic measuring techniques it is advantageous to use ferrite cores instead of metal sheets due to the simpler applications. The ferrites are also usable at higher frequencies, up to 150 kHz. A minor drawback of the ferrites, though, is their brittleness. However, in most applications, this causes no serious problems.

The equivalent circuit diagram of a magnetostrictive transducer is relatively problematical because of the non-linear behaviour of the transducer, and the expression for the losses. High power values are usually not used in the measurement technique, so that this enables the use of a simplified equivalent diagram as given in Fig. 3.9. It is analogous to the equivalent circuit diagram of the piezoelectric transducer. The parame-

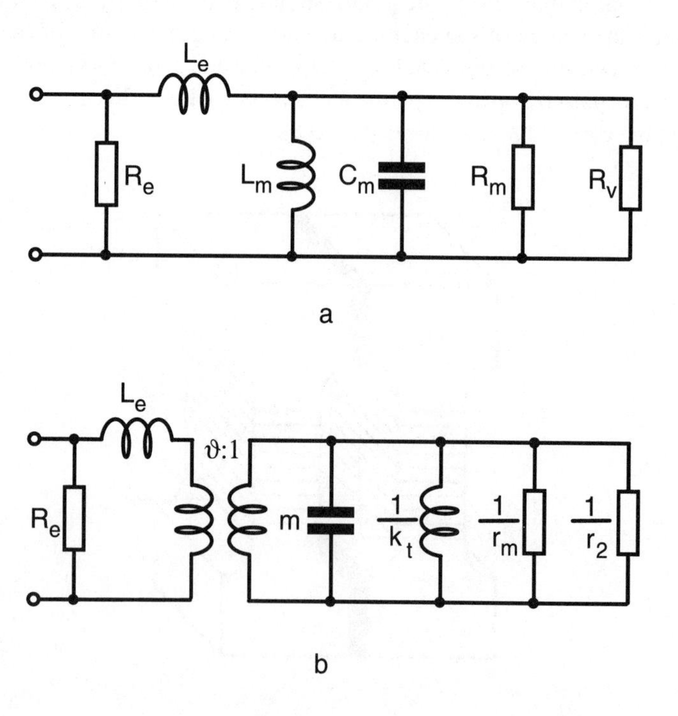

Fig. 3.9 ***An equivalent circuit of a magnetostriction transducer, a) electrical, b) electro-mechanical***

ters, L_e and R_e, in the electrical equivalent circuit represent the inductance and the electrical resistance of a fully loaded transducer while the values of the equivalent inductance, L_m, the capacitance, C_m, and the resistance, R_m, depend on the mechanical properties of the transducer.

$L_m = \gamma^2/k_t$	is inversely proportional to the stiffness, k_t
$C_m = m/\gamma^2$	is a dynamical mass, m
$R_m = \gamma^2/v_m$	are the mechanical losses of the transducer
$R_v = \gamma^2[(\rho_1 C_1 + \rho_2 C_2)S]$	represents the acoustic resistance of the adjacent media

(3.17).

and the transformation coefficient is given by the expression

$$\gamma = \frac{4\pi}{10^4} \cdot \frac{L\lambda}{l.n} \tag{3.18}$$

where L - inductance of the system neglecting the hysteresis losses caused by eddy currents

λ - magnetostriction pressure coefficient $[N.Wb^{-1}]$

l - length of the mean magnetic line of force [m]

n - number of windings per unit length of the mean magnetic line of force.

The parameters of the equivalent circuit can be determined from the circular impedance diagram.

3.4 Transducers operating with an electric discharge

3.4.1 Introduction, classification and features of a discharge

Of the many kinds of electric discharge that exist, an electric spark and a glow discharge can be used for electroacoustic transducers. In practice, both kinds of discharge can exist at *normal* atmospheric pressure, p (around 100 kPa), in a *non-homogenous* field – between the tips of the electrodes. They are classified as short discharges with an intensive contact at the electrode. A *short discharge* means that the so called discharge path (distance d, between the electrodes), is small (0.2 – 1.5 mm).

An electric spark creates a thermal source of wave pulses (ultrasonic or audible) for the measurement technique, with the following properties:

a) the possibility of creating a source with minimum size, i.e., an almost point-shaped source, which leads to an omnidirectional radiation pattern, and a spherical wave
b) large acoustic power in transmitter mode, a wide frequency range, and a steep leading edge
c) random character of the spark trajectory, which results in an uncertainty in the position of the wave source whereby the intensity of the generated sound pulse fluctuates as well
d) strong electromagnetic interference in the surrounding space
e) unusable in explosive surroundings.

A glow discharge has parameters that depend on pressure changes (including acoustic pressure changes), which can be evaluated electrically. It is possible to exploit this effect in a discharge microphone as a receiver of ultrasonic waves, with the following features:

a) the possibility of creating a compact receiver
b) the microphone has no moving parts and thus has a high upper frequency limit
c) it enables the design of a reciprocal transducer
d) it requires the use of a HV power supply (kV range)
e) it can only be used in a non-explosive medium
f) the instability of the discharge causes interference of its own.

When using sparks, the drawbacks can be overcome to such a large extent that the spark method is slowly becoming a suitable source of ultrasonic waves for practical use.

In connection with a discharge microphone, the use of a glow discharge is possible only with some limitations.

3.4.2 The electric spark – a source of ultrasonic waves

The description of a spark is given by two theories.

a) The Townsend theory assumes the creation of a conducting channel in two phases:

1 creation of an avalanche - the creation of groups of ions and electrons
2 discharge - upon increasing the voltage, the number of avalanches and ions increases too until a discharge (a spark) arises.

It follows from the Townsend theory that there is a certain dependence of the breakdown voltage on the spark gap distance. A limitation on the validity of the theory is given by an inequality [3.6],

$$p \, . \, d \leq 266 \qquad [\text{Pa.m}].$$

For $p = 101.325$ kPa, this yields $d \leq 2.6$ mm.

b) A channel discharge (Striegel, Loeb, Meek) assumes the creation of streamers, i.e. when the avalanches build up a conducting channel – a streamer, by their mutual connection. The build up of a spark occurs very rapidly in this way. The event is highly intensive thermally, and thus very audible as well. In practice, the channel discharge can be created by increasing the breakdown voltage, U_p, which can be determined from the Townsend theory. A minimum necessary increase in the voltage is given in [3.6] for values of the product p.d:

p.d [Pa.m]	500	250	200	100	0
increase [%]	10	14	18	22	25

Thus, from the Townsend theory, for the maximum possible distance of 2.6 mm, $U_p = 10.8$ kV. An increase of 14 % corresponds to 1.51 kV. So the total voltage necessary for a channel discharge is

$$U_{pchan} \geq 12.31 \text{ kV}.$$

This is one of the requirements for the voltage supply.

The small distance between the tips in a spark gap, together with their finite dimensions, means that the spark gap system does not behave as an ideal system with point tips at a large distance, and thus the dependence of U_p on distance d is non-linear.

3.4.3 Voltage supply for the electric spark

Three types of supply are available:

a)a constant voltage source with the voltage higher than U_p, connected by a HV switch

b) a pulsed source - e.g. a transformer

c) a piezoceramic source.

The optimal excitation is performed by a pulse transformer. The principal circuit diagram is shown in Fig. 3.10.

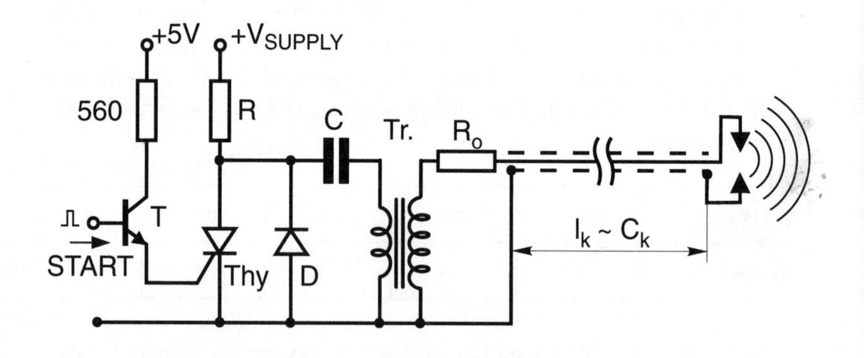

Fig. 3.10 ***A circuit for the excitation of a spark gap by a pulse transformer***

The energy of a capacitor is discharged into the primary winding of a transformer. From the secondary winding, the high voltage is drawn, and it increases in time, approximately as sinωt. The instantaneous voltage, u_p, at the moment of breakdown is always higher than the static breakdown voltage, U_p, because it continues increasing during the building-up of the spark. In this way, the channel mechanism of a discharge is secured.

The electrodes are pointed, so that the spark is positioned more precisely in space. A coaxial cable of length l_k, and capacitance C_k, serves as a charging line. Its energy, $E = (1/2)\, C_k u_p^2$, discharged between the tips is high enough to create an acoustic wave, and the resistor, R_o, at the beginning of the line rejects interference efficiently.

3.4.4 Spark features

The V-A characteristics of a spark for the excitation system described above is illustrated in Fig. 3.11. A coaxial cable serves as a connection in this particular case, with a capacitance $C_k \cong 180$ pF, $R_o \cong 4.4$ kΩ, and the spark gap distance was d = 0.4 mm.

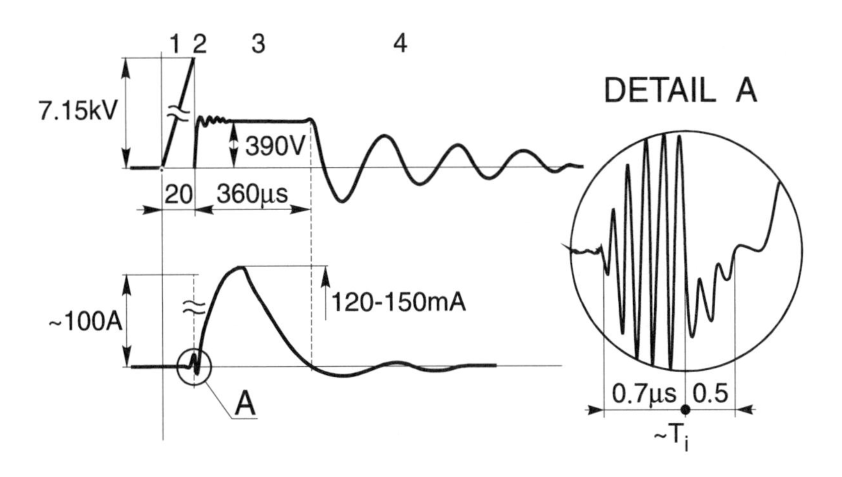

Fig. 3.11 ***V-A characteristic of the spark gap during pulse excitation***

The time course can be divided into several parts:

1 - increase of the supply voltage until breakdown
2 - breakdown - the voltage drops rapidly, and the current reaches high values. In this region, the acoustic signal is generated
3 - glow discharge - no influence on the acoustic signal
4 - attenuated oscillation - energy is dissipated by oscillations.

Because of the rapid discharge of energy, the channel between the tips (inside the spark gap) is intensely heated, the pressure increases, and the gas expands into the space surrounding the discharge. At the beginning the pressure propagates as a shock wave, then afterwards as a sound wave. At first, the shape of the wave is cylindrical, but already within a small distance from the axis, it changes to a spherical wave.

3.4.5 A spark – a thermal source of sound

A spark as a thermal source of sound can be described mathematically. Utilising knowledge from thermodynamics and acoustics and by a proper mathematical treatment, equations can be obtained which describe the qualities of the sound wave transmitted at the spark site (the channel surface), namely
a) the pressure conditions (3.19)
b) the frequency spectrum (3.20).

A full mathematical treatment is given in [3.6].

$$p_{(t)} = \frac{(\varkappa - 1)q}{\pi r d c} e^{-\frac{c}{r}t} \sin\left(\frac{c}{r}t\right)$$

for $t \in \langle 0, T_i \rangle$, and (3.19)

$$p_{(t)} = p_{(t - T_i)} \quad \text{for } t \in \langle T_i, \infty \rangle$$

$$F(j\omega) = K \frac{2\left|\sin\left(\frac{T_j \omega}{2}\right)\right|}{\sqrt{4c^4 + r^4\omega^4}} e^{j\left(\arctan\frac{\sin(\omega T_j)}{1 - \cos(\omega T_j)} + \arctan\frac{2\omega c r}{2c^2 - \omega^2 r^2}\right)}$$

(3.20)

where $K = \frac{(\varkappa - 1)q}{\pi d}$;

T_i - spark duration
c - velocity of sound propagation
r - radius of the spark channel
q - thermal power in channel
d - distance between tips.

The thermal power supplied depends on the spark gap distance, d, and, this influences the radius of the spark channel, as well as the time T_i.

From a plot of equation (3.20) one can infer these parts of the spectrum that are radiated with the highest intensity, and how they are related to the spark parameters. For example:

At the primary winding of the pulse transformer, C = 1 μF, U = 300 V, (energy 45 mJ). The open-circuit output voltage (negative with respect to the framework) increases to 26 kV, in 60 μs. For a capacitance of the connecting cable of 180 pF, and with d = 0.4 mm, these give U_{break} = 7.15 kV, a discharge energy 4.6 mJ, T_i = 1.22 μs, and r = 0.18 mm; the maximum in the frequency spectrum occurs at 300 kHz [3.4].

3.4.6 Transducer design

The design must satisfy some requirements, the most important of which are:

a) a reliable grip of the electrodes, perfect alignment, and the capability of adjusting the gap
b) good and reliable insulation of both electrodes and leads
c) a minimum of acoustic shadowing
d) resistance to wearing down.

Two ways are possible to achieve these requirements:

1) with a discharge in air (Fig. 3.12a)
2) with a discharge on the boundary between air and a dielectric (Fig. 3.12b).

Among suitable electrode materials, tungsten seems to have the best qualities, and with this material the transducer properties do not change significantly after 10^4 - 10^5 measurements have been carried out. Then the gap increases, and as a consequence, the acoustic power increases as well, and the frequency spectrum shifts towards the lower frequencies.

The spark parameters depend on the build-up of conductive channels and these have a random variance around the statistical mean value. For small d changes in the channel thickness dominate, whilst for larger values of d it is the misalignment. A change in these parameters influences the measurement precision because the position of the source of the sound wave varies while the position of the transducer remains constant. A recommendable choice of d thus lies within the interval of 0.3 - 0.5 mm.

3.4.7 Safety of operation and conclusions regarding sparks

It is not sensible to use an electric spark in explosive surroundings. Regarding the design and the parameters of the power supply, the safety of

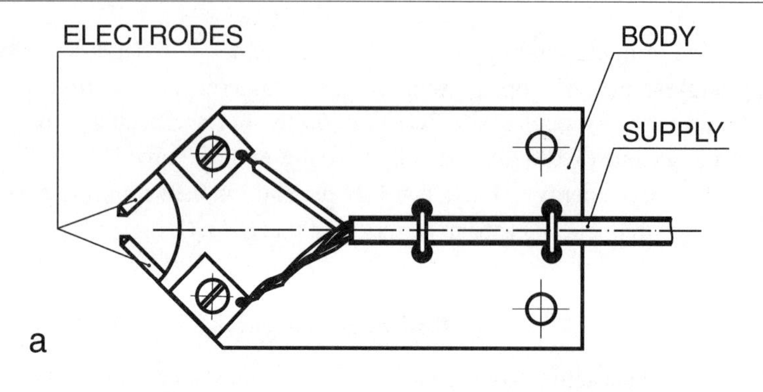

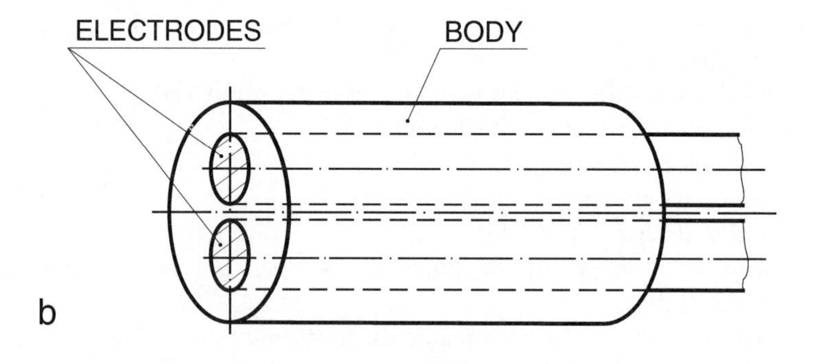

Fig. 3.12 ***A schematic drawing of a spark gap for generation of a spherical ultrasonic wave pulse***

the operational staff is maintained. However, unpleasant shocks may occur on direct contact with the live electrode. The live electrode, though, does not tend to flashover onto its immediate neighbours, unless the shielding of the leading cable is impaired, which is improbable.

An electric spark is almost a point source of sound waves with large acoustic power and with the maximum of its frequency band shifted to ultrasonic frequencies. It produces a steep leading edge to the acoustic pulse which is beneficial to the precision of pulse detection.

The minimum error caused by random variations of the spark displacement as the source of acoustic pulses reaches a value of about ±0.1 mm.

As suitable receivers for the detection chain, capacitor microphones, described earlier, are satisfactory.

3.4.8 A glow-discharge as a receiver of ultrasonic waves

A glow discharge is a stable, self-contained discharge defined in the current region between about 10^{-5} - 1 A. Further characteristic features include cold electrodes, and an alternation of light and dark bands along the discharge path.

The predominant utilization of this discharge is in the low pressure region (10^2 - 10^3 Pa). On increasing the pressure to atmospheric level (~101 kPa), a narrowing of the discharge appears, and the so called cathodic space shortens considerably. The discharge parameters are pressure-dependent. This also means that an acoustic pressure, which in fact may be produced by a sound wave, may act as a pressure change and can be detected electrically.

It can be shown theoretically [3.4] that a discharge current is pressure-dependent. Current changes following amplification provide information about the incidence of a sound wave upon a glow discharge.

3.4.9 The principle of the discharge microphone

Again, the discharge occurs between two pointed electrodes, so that it is defined in space with a relatively high precision. The power supply and part of the evaluation circuit is shown in Fig. 3.13.

R_f and C_f filter the supply voltage V_s. The current is limited by the resistor R_t, so that it remains in the glow-discharge region. Between the tips, a voltage V_d arises from the glowing discharge. The discharge current also passes through the resistor R_k, from whence the voltage signal (the voltage change) is picked up, via the protective circuit (R_o,D_1,D_2), and the separation circuit (C_v), into the amplifier. The second way of obtaining a signal is by positioning a small probe - an aerial - close to the tips. Through the resultant capacitive coupling, the change in the signal (after the incidence of an ultrasonic wave) is led directly to the amplifier input. The signal-to-noise ratio increases, and the problems with separation of the high bias voltage disappears.

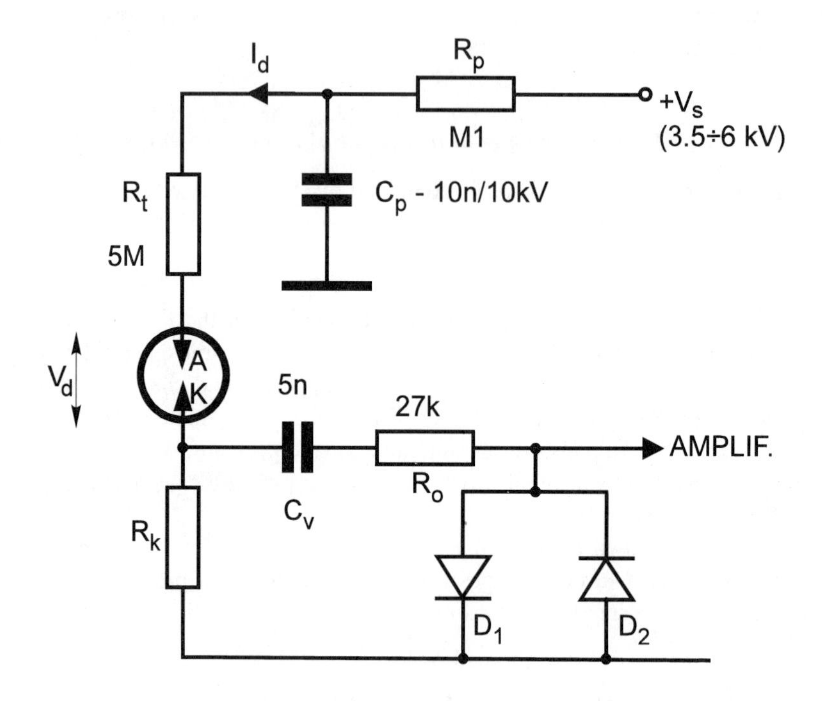

Fig. 3.13 ***Power supply and detection circuits for a discharge microphone***

The value of R_t lies in an interval between about 2 - 10 MΩ. The sensitivity depends on the distance between the tips; the maximum value lies at around 1 mm.

The material of the anode (A) influences the stability of the discharge. A material with good thermal conductivity, e.g. copper, is recommended. The material of the cathode (K) influences the sensitivity, noise, etc. A suitable material is tungsten. A sharp tip suits the shape of (K); the shape of (A) should be slightly rounded [3.4].

3.4.9.1 The principle of the reciprocal transducer

Using a discharge microphone offers the opportunity of using the same transducer as an emitting-receiving transducer (ERT).

In Fig. 3.14, a circuit design is shown which illustrates the combination of the two principles described above. The left part of the circuit diagram represents a discharge microphone circuit, the right part is the circuit of a spark transmitter with pulse excitation. The spark is connected through an auxiliary spark gap ASG, sound-insulated from the surrounding medium. The gap distance must be larger than in the ERT, so that a glow-discharge cannot arise. The voltage from Tr must be sufficiently high to cause a breakdown over the spark gap distance, increased and doubled in this way. 1 and 2 are the coaxial cables functioning as voltage leads to the system. The ASG must be close to the ERT. The coaxial cable, 1, along with the resistor, R_f, create a filter for the microphone supply voltage. The signal for the amplifier is obtained using the probe S described above. The microphone is in permanent readiness for reception except during the instant of transmission (during a spark) and then during the following interference of its own (a duration of about 1 - 1.5 ms).

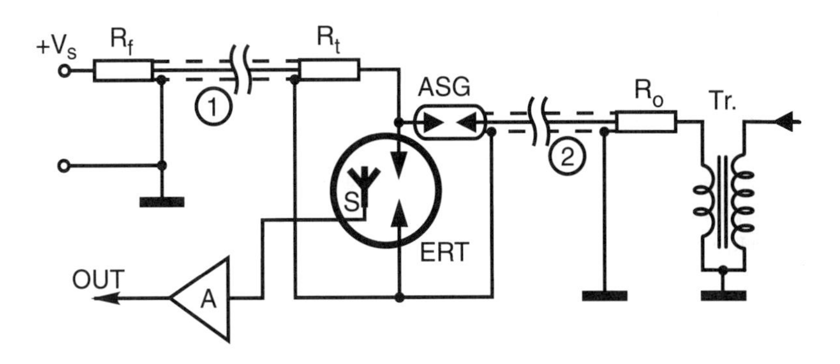

Fig. 3.14 ***A combined circuit for the excitation of a spark and the detection by a plasma discharge of an ultrasonic wave pulse***

So far the time unstability of the discharge (the self-oscillation) has been regarded as a drawback of the discharge microphone. Furthermore, it leads to fluctuations in some of the parameters, especially the sensitivity, and in the discharge position in space. Amongst the construction and design problems, a high supply voltage (of the order of kV) and the creation of ozone within the discharge may have a corrosive influence on the surrounding electrical parts and elements.

3.5 References

3.1 *Boys J. T., Strelow E. R., Clark G. R. S.:* A prosthetic aid for a developing blind child. Ultrasonics, No. 1, Jan. 1979, pp. 37-42

3.2 *Biber C. et al:* The Polaroid Ultrasonic Ranging System, Preprint, Proc. 67th Convention. of the Audio Eng. Soc., New York, Oct.31-Nov. 3., 1980 Audio Eng. Soc. Publ. Co.

3.3 *Merhaut J.:* Teoretické základy elektroakustiky, Academia, Prague, 1985 (in Czech)[1]

3.4 *Toman M.:* Generovanie a snímanie ultrazvukových impulzov pre meranie priestorových súradníc, Thesis, Electrotechnical Faculty, Slovak Technical University, Bratislava, 1990 (in Slovak)[2]

3.5 *Kikuchi Y.:* Ultrasonic transducers, Corona Publ. Co. Ltd, Tokyo, 1969

3.6 *Warren M.P.:* Physical Acoustics - Principles and Methods, Academic Press, New York, London, 1964

3.7 *Woollett R. S., Le Blanc L.:* Ferroelectric Nonlinearities in Transducer Ceramics, IEEE Transactions on Sonics and Ultrasonics, SU-20, No. 1, 1973

3.8 *Lerch R.:* Messung piezoelektrischer und elastischer Größen von Piezopolymer-folien für elektroakustische Wandler, Technisches Messen 48, Jahrgang 1981, Heft A, 287-294

3.9 *Lancee O. T. et al:* Transducers in Medical Ultrasound: Part 1 - Ferroelectric Ceramics versus Polymer Piezoelectric Materials, Ultrasonics 23, No. 3, 1985, 138-142

3.10 *Veverka A.:* Technika vysokých napětí, SNTL, Prague, 1982 (in Czech)[3]

3.11 *Granovskii V. L.:* Elektricheskii tok v gaze, ustanovivchiisya tok, Nauka, Moscow, 1971 (in Russian)[4]

---

[1]Theoretical principles of electroacoustics

[2]Generation and sensing of ultrasonic pulses for measuring spatial coordinates

[3]High voltage techniques

[4]Electric current in a gas; a stationary current

4 Ultrasonic digital measuring methods

An advantage of measuring methods exploiting ultrasonic waves is that they enable the direct accomplishment of a digital measurement without the need for conversion of an analog signal to a digital one by an analog-to-digital converter (ADC). In most applications, either the velocity of the ultrasonic wave or the time of flight of a wave over the measured distance is utilized. From a technological point of view, another advantage is that the propagation velocity of ultrasonic waves is many orders lower than that of electromagnetic waves. Because of this, less stringent demands are required of the transducers and the electronic circuits. On the other hand, the use of ultrasonic waves has certain drawbacks too, in particular the large dependence of the velocity of the ultrasonic waves on the parameters of the medium, and their high attenuation and scattering, especially in air.

These circumstances determine the boundaries for the full exploitation of ultrasonic measuring methods. In the first instance, they are limited to relatively small distances, compared with methods employing electromagnetic radiation. In cases where a more precise measurement is required, ultrasonic techniques become more complicated due to the need to compensate for the effects of fluctuations of the parameters of the medium. It is partly possible to diminish the scattering of ultrasonic radiation using matrix transducers, especially in medicine.

In the succeeding sections of this chapter some general methods are described which are most often used in coordinate and dimensional measurement, especially in air. Here the problems of measurement arise in the most significant way. In later sections, measuring methods for some particular applications are described.

4.1 Fundamental ultrasonic measuring methods

In most technical applications, the velocity of propagation of an ultrasonic wave is utilized for the evaluation of the measured quantity. The elementary physical properties and relationships for the propagation of ultrasonic wave motion have already been described in Chapter 2. Expressions for the propagation velocity in all kinds of media are presented there; however, only an approximate relation is given, (2.19), for the temperature dependence of the velocity of ultrasound. In general, this dependence is non-linear and it is different for particular materials. Moreover the velocity of ultrasound propagation also depends on other parameters of the medium, e.g. humidity, air flow, etc.

The measuring methods are determined to a considerable extent by the necessity to reject these effects. Therefore in the next section attention is devoted to a more detailed examination, and to a quantitative evaluation, of this phenomenon. This is useful since an effective way of increasing the precision of a measurement is by the correction of the measured readings, based on the known physical dependence of the propagation velocity on the parameters of the medium.

The influence of the physical parameters of a medium upon the propagation velocity of an ultrasonic wave shows up most strongly in the case of air. Therefore a more detailed examination is devoted to this particular problem. From Boyle's law, it follows that

$$\frac{p_a}{\rho} = \frac{R\Theta}{M} \tag{4.1}$$

where M - molecular weight
Θ - absolute temperature [K]
R - universal gas constant.

Substituting (4.1) into (2.17), the velocity of sound can then be expressed as

$$c = \sqrt{\varkappa \frac{R}{M} \Theta} \tag{4.2}$$

where $\varkappa$ - Poisson's constant.

After introducing the expression for the sound velocity c_o, at a temperature of 0 °C,

$$c_o = \sqrt{\varkappa \frac{p_0}{\rho_0}} \qquad (4.3)$$

where p_o - atmospheric pressure at 0 °C
ρ_o - air density at 0 °C,

expression (2.17) can be rearranged as

$$c = c_o \sqrt{\frac{\Theta}{\Theta_0}} = c_0 \sqrt{\frac{\Theta_0 + \vartheta}{\Theta_0}} = 331.56\ \sqrt{1 + 0.366\vartheta} \qquad (4.4a)$$

where $\vartheta = \Theta - \Theta_0$ - temperature in °C.

Its linearized form is

$$c \cong c_o \left(1 + \frac{\vartheta}{2\Theta_0}\right) = 331.56\ (1 + 1.831 \times 10^{-3} \vartheta) \qquad (4.4b).$$

A further approximation also includes a quadratic term

$$c = 331.56 + 0.6085\vartheta - 0.00049\vartheta^2 \qquad (4.5).$$

The influence of temperature represents an error in a measurement of 0.1381 %/°C, and it is thus relatively significant, especially if more precise measurements are required.

The influence of air humidity is smaller. In Chapter 2 a linear dependence was presented. In fact the dependence is rather non-linear, especially up to 20 % of relative humidity. In [4.3], a more detailed examination of this dependence on humidity, and frequency, is given. In [4.1], the following expression is introduced

$$c_d = c_h \sqrt{1 - \frac{e}{p_a}\left(\frac{\gamma_w}{\gamma_a} - \frac{5}{8}\right)} \qquad (4.7)$$

where c_d - velocity of sound in dry air
c_h - velocity of sound in humid air

e - pressure of water vapour
p_a - atmospheric pressure
γ_w - the ratio of specific heats at a constant pressure and a constant volume for water vapour
γ_a - the ratio of specific heats at a constant pressure and a constant volume for air.

The air flow represents a further significant contribution to the velocity of propagation of an ultrasonic wave. The measurement error caused by the air flow is proportional to the velocity vector in the direction between the transmitter and the receiver. This influence is described in more detail in Chapter 6. Here it should be pointed out that a more significant error can also arise in laboratory conditions, if a more accurate measurement is required. For example, the change of propagation velocity is already 0.15 %, for an air flow velocity of $v_a = 0.5$ m/s (in the direction of sound propagation). In most cases this error adds to the resultant measurement error. Thus in a measurement of a distance of 1 m, the air flow velocity causes an error of 1.5 mm.

Rejection of the undesirable effects caused by the physical parameters of air mentioned above can be achieved in several ways, all of which affect the measuring methods to a considerable extent. If the value of a parameter and the dependence of the ultrasonic wave velocity on it are known (so called parametrical sensitivity), a compensation, or correction, can be introduced.

If the detector for a particular physical quantity yields an electrical output signal, this can be passed to the input of an electronic evaluation circuit, and according to the known physical dependence, it can act upon the characteristics of the partial circuit in question, thus lowering or even compensating fully the influence of the side effect upon the detected physical quantity. This can easily be performed, for instance, for temperature, because the temperature is detected by electrical sensors and the dependence of the ultrasonic wave velocity on temperature is linear (to a large extent). This can be achieved by using a voltage-controlled oscillator or by a voltage-controlled amplifier. A more difficult case is that of introducing a compensation for the influence of air humidity. This has a non-linear effect on the propagation velocity of an ultrasonic wave and thus needs a more complicated detector. The most difficult compensation is that for air flow. This is a vector quantity, and a yet more complex detector arrangement is necessary for its measurement.

For these reasons, only temperature compensation is used in most cases. If the detector has a digital output, a correction to the measured value can be introduced by use of an algorithm. However, it needs digital processing of the particular physical quantity, and knowledge of the physical dependence of the sound velocity upon it.

The increase in the precision of a measurement by digital correction of the measured values is very significant. However, it requires auxiliary instrumentation for the measuring instrument, by adding multi-parameter sensors. These probes should be placed in the measurement area.

With the development of computer technology, digital measuring methods have become widespread. They are based on the principle of comparing the measured quantity with a known value. Such comparative methods enable simultaneous compensation for several quantities, without the need for their direct measurement. These no doubt require more complicated (even multiple) measuring circuits, but they obviate the need for more sensors. They also enable a further increase in the precision of measurement. These methods are described in more detail in the following chapters, especially in Chapter 6.

Ultrasonic measuring methods have developed in parallel with other methods, especially with those which use electromagnetic waves. Therefore they are similar (up to a point). However, there are differences. The differences result from fundamental principles and from the different properties of ultrasonic wave motion. This was mentioned in Chapter 2. From the point of view of the differences, the most important one is that ultrasonic waves can only propagate through a material medium - solid, liquid or gaseous. A considerably lower velocity of propagation of ultrasound results, compared with electromagnetic waves (by 5 to 6 orders of magnitude), depending on the particular medium.

This essential feature, that ultrasonic wave motion is transferred by a medium, causes its dependence on the state of the medium. The medium influences the velocity, the attenuation and the scattering of the waves, all of which limit the precision and the operating range of measurement.

On the other hand, the lower velocity of propagation of ultrasound allows the use of short-distance pulse measurements, as well as the use of some other methods. Also the electronic circuits are simpler and less demanding.

The measuring methods can be roughly divided into two fundamental groups. One of them uses a continuous (harmonic, periodic) wave motion, the other one uses discontinuous (pulsed) waves. The case of harmonic ultrasonic waves is commonly used to evaluate the phase shift between the transmitted and the received signals. For this, the phase change is dependent on the measured quantity, e.g. a distance, in an unchanged and stationary medium; on the composition of the (fluid) medium at a stationary distance between the transducers; and on the medium flow rate, temperature, etc. Therefore, these methods can be included in the group of methods which use phase modulation.

In the case of measurement using a pulsed ultrasonic wave, the duration of the pulse propagation, from the transmitter (pulse generator) to the receiver (pulse front detector), is usually evaluated. The duration of the pulse propagation depends on the transmitter-receiver distance, as well as on the properties (and the motion) of the medium intervening. Generally this method can be classified among those using time-pulse modulation. In the case of the detection of a reflected pulsed ultrasonic wave from an object, the method is analogous to the radiolocation method. In the literature, this is sometimes called an echolocation method.

Both methods have advantages and drawbacks of their own. These will be described and compared later. In order to exploit the advantages and overcome the drawbacks, various solutions have been developed. In some of these, elements of a combination of both methods can be found.

4.1.1 Measuring methods using phase evaluation

For an explanation of the principle and features of this method, we use the instructive case of measurement of the distance between a transmitter and a receiver. This is a case of measuring a position or a movement in a straight line (Fig. 4.1a). The emitting transducer is excited by a generator of a harmonic voltage with a constant frequency, f. The transmitter radiates harmonic ultrasonic waves which fall on the active surface of a receiver at a distance x from the active surface of the transmitter. The phase of the amplifier output voltage, V_a, is compared in the phase detector with the phase of the generator voltage, V_g. The output voltage, V_p, of the phase detector is led to a comparator where it is shaped to a rectangular (digital) pulse, V_c. The number of pulses, V_c, is counted by a digital counter. At its output, a parallel combination of signals appears whose logic level corresponds to a number N_x, in a suitable prescribed code.

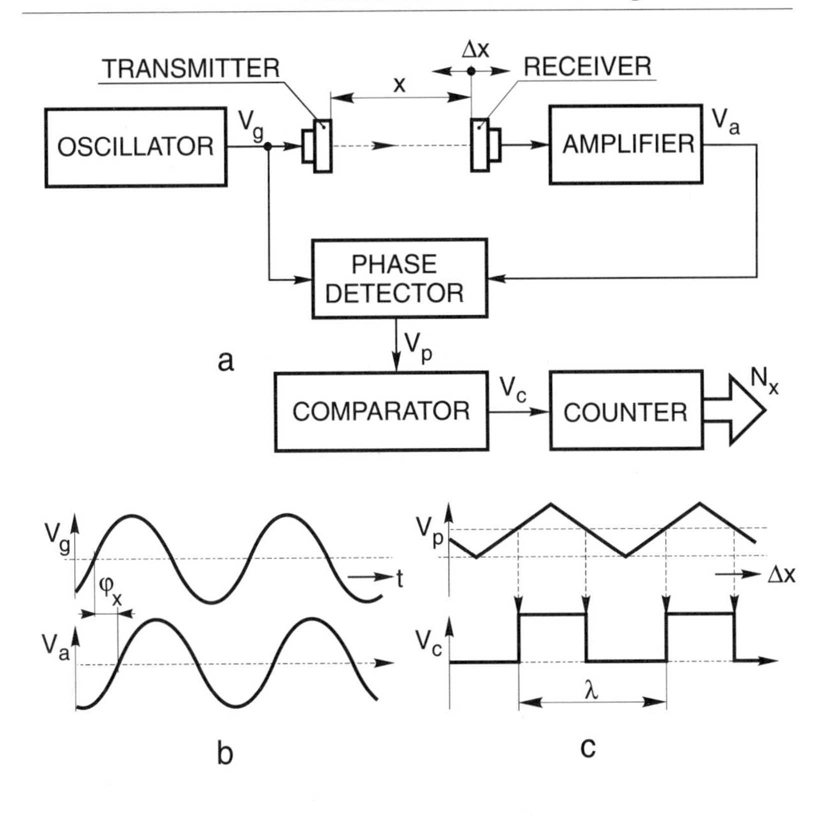

Fig. 4.1 ***Measurement of distance by the phase method***

Let us start with the assumption that the receiver is at rest, idling, without movement, at a distance x from the transmitter. Furthermore, let us assume a certain phase shift, φ_x, between the voltages V_g and V_a and a zero state of the counter (Fig. 4.1.b). The counter state gives the initial (starting) position of the receiver which is mechanically connected to the located object. From this position, the motion of the receiver, Δx, is measured (Fig. 4.1.c). It is a case of an incremental measurement in which one wavelength corresponds to one increment. At the same time, this increment represents the resolution of the method.

Except for the first change of contents of the counter, its state will alter after each movement of the receiver position by a path equal to one wavelength $\lambda = c/f$ (2.1). The path traversed is expressed as a multiple of wavelengths

$$\Delta x = (N_x \pm 1)\lambda \qquad (4.8).$$

Since the initial position of the receiver can, in general, be arbitrary within a single wavelength, an error can arise on evaluation of the trajectory which can reach a value close to one wavelength. It is a discretization error. In order to reduce this, a circuit can be used which converts a continuous change of the phase detector output voltage within one wavelength into a number (both at the start and at the end of counting λ), and after the measurement is over, the measured value is enhanced by this lower-order supplement. The calculation of this supplementary number can be performed by a simple logical consideration derived from Fig. 4.1.c.

In the block circuit diagram shown in Fig. 4.1, the counter cannot distinguish the direction in which the receiver moves, and thus, neither the direction of movement of the located object. In fact, only the total path change of the receiver is measured, from which not even its absolute position relative to the inital displacement can be determined. In order to sense the position during bidirectional movement of the receiver, the direction of the motion must be distinguished and a reversible counter must be used. A circuit which distinguishes the direction of the motion can be created by adding one more phase detector to the electronic circuitry, with a voltage V_g, shifted by 90° in phase, and connected to its input. It provides one more position-shifted voltage signal by means of which the evaluating logic circuit can distinguish the direction of motion.

From the point of view of precision, especially that of resolution in this actual case (the discretization error), a wavelength as small as possible is desirable, i.e. as high as possible a frequency of the voltage V_g. However, with increasing frequency of sound waves, their attenuation increases as well (especially in air). For example, at a frequency of 10 kHz, the drop of the acoustic signal level is 1 dB on increasing the distance p_o by every 500 mm, while at a frequency of 1 MHz, the same drop of 1 dB arises every 5 mm of distance increase. The contrast between the requirements of a high measurement precision and a wide measurement range can be efficiently solved in an electronic way - by introducing a frequency divider and a frequency multiplier into the measuring circuit (Fig. 4.2). An arrangement for distinguishing the direction of the motion is also illustrated in that circuit.

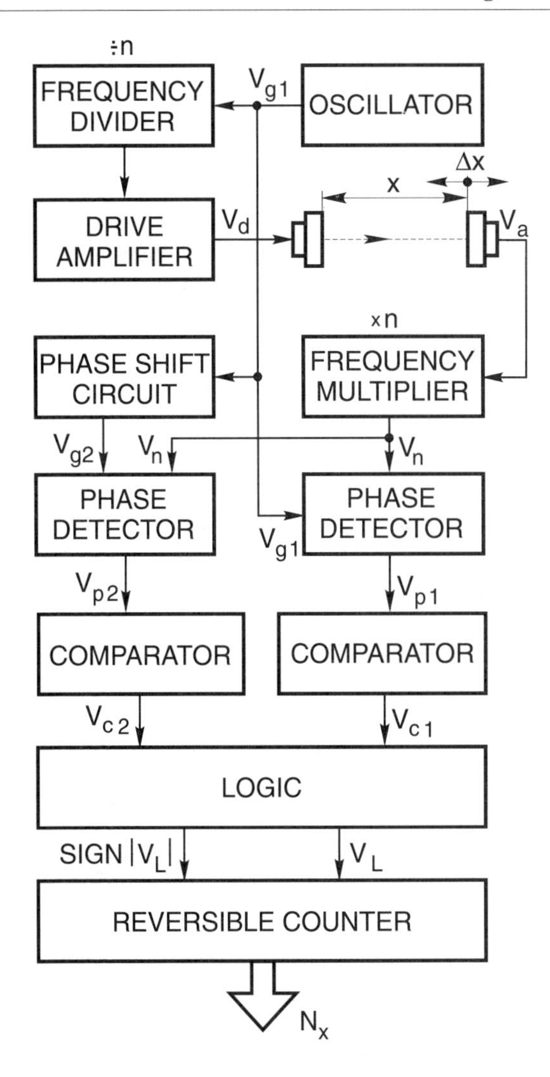

Fig. 4.2 ***Measurement of distance by the phase method with increased resolution and detection of the direction of movement***

Following the harmonic voltage generator, a frequency divider is introduced, through which the frequency of the voltage, V_d, is lowered n-fold. In this way, the frequency of the ultrasonic wave is lowered too, whereas the electronic circuit works with the original, higher, frequency. If an acceptably low frequency of the ultrasonic wave is maintained, the resolution increases n-fold. For correct operation of the electronic circuit, a frequency multiplier must be inserted into the circuit chain so that the frequency of the voltage V_g equals the frequency of the voltage V_a.

By this extension of the circuit, an acceptable attenuation of the ultrasonic signal is achieved whilst simultaneously increasing the resolution. If, for example, a frequency of the exciting voltage V_d of $f = 34$ kHz is selected, it corresponds to a wavelength of $\lambda \cong 1$ cm. Using a frequency divider and a multiplier with $n = 100$, the frequency of the generator becomes $f_g = 3.4$ MHz. The voltage V_n has the same frequency. The voltage V_p, at the phase discriminator output, corresponds to a 100-fold lower virtual wavelength, i.e. about 0.1 mm.

By introducing a second channel with an input voltage V_{g2}, shifted by 90° to V_{g1}, distinguishing the direction of movement can be achieved using a logic evaluation circuit and, moreover, a four-fold increase in the pulse counts (4 pulses per one virtual wavelength). In such fashion a change of $\lambda/4$ (i.e. 25 μm) corresponds to a single pulse, V_L, which approaches an acceptable value.

4.1.2 Pulse measuring methods

Pulse methods evaluate the time of propagation of an ultrasonic pulse wave from a transmitter to a receiver. For the case of distance measurement, a typical example of a digital evaluation of the propagation time is illustrated in Fig. 4.3a.

At the start of a measurement, a pulse generator excites a voltage pulse, V_i. This is converted in the transmitter to an ultrasonic pulse which propagates towards the receiver at a velocity c. After the front edge of the pulse wave falls on the active surface of the receiver, a voltage V_a appears at the amplifier output which ends the time measuring pulse, V_p. The conversion of the pulse propagation time into a number N_x is then carried out by a circuit depicted in the lower part of Fig. 4.3a. At its input there is a bistable circuit whose output is set to the active (logical one) state. This results in the opening of a gate through which pulses V_c begin to pass from the clock (time mark) generator to a reversible counter. After arrival of the pulse V_a, the bistable returns to its initial (zero) state, and the measurement is over. On the counter display, a number N_x ap-

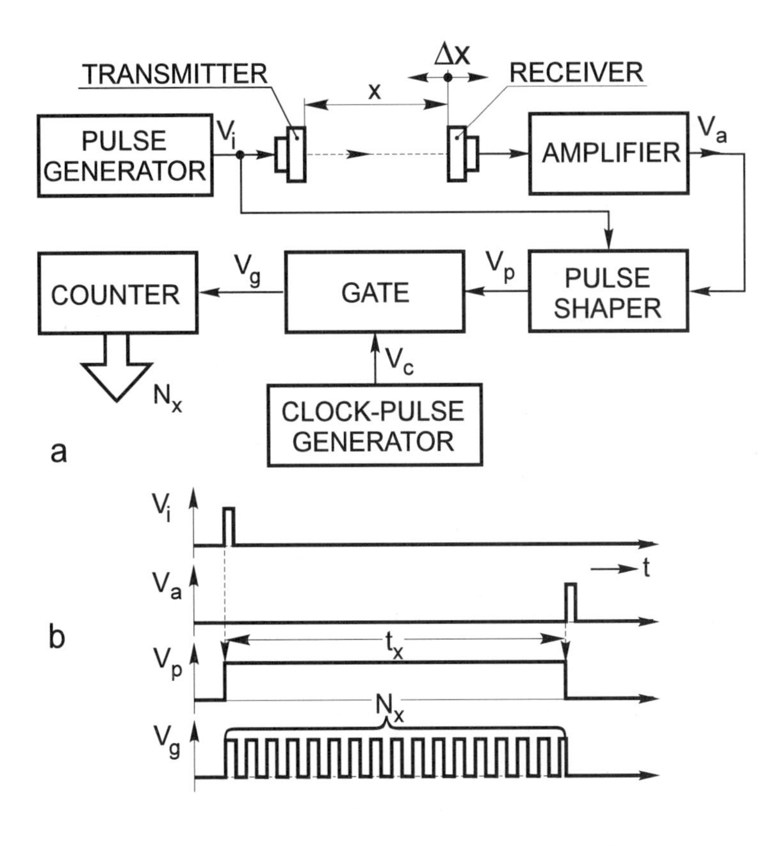

Fig. 4.3 ***Measurement of distance by pulse method***

pears which corresponds to the measured distance. The process described is illustrated in time diagrams in Fig. 4.3b.

The measured distance, x, can be expressed by the propagation time, t_x, of the ultrasonic wave as

$$x = ct_x .$$

At the same time, it is also true that

$$N_x = ft_x .$$

Allowing for the error of the time-to-number conversion (a discretization error), one may write

$$x = (c/f) \, . \, (N_x \pm 1) = k_t \, (N_x \pm 1) \qquad (4.9).$$

It should be noted that the conversion coefficient $k_t = c/f$ has a dimension [m], which corresponds with the dimension of the measured distance. In order to read the distance in a straightforward way in length units, its value should be expressed in decimal order of tens. The indication in the preselected units is achieved by suitable placement of the decimal point. Then the frequency value reading is not rounded off. If using a crystal oscillator as a clock (time mark) generator, its frequency must be finely adjusted according to the velocity of the ultrasonic wave at the selected conditions of measurement.

In order to lower the discretization error, the working frequency is usually chosen sufficiently high. As a rule, tens of MHz are used. For example, if 20 °C is the ambient temperature in the working space, the frequency can be 34.37 MHz. On measuring a distance of $x = 1$ m, there is a reading N_x of 100 000 ± 1 on the counter display. The resolution is then 10 μm.

To increase the precision of the measurement several methods can be used. In the simplest case, correction of the measured value for the actual temperature (and, incidentally, humidity too), by a simple calculation, can be performed. Another possibility is through fine frequency tuning of the clock pulse generator. This can be accomplished, for instance, so that the receiver is placed at a suitable (known) distance and the clock generator is adjusted to a value at which the reading on the display equals the preset distance. However, this method is rather variable because adjustable clock pulse generators have, as a rule, a lower frequency stability. Moreover, the parameters of the ambient atmosphere can sometimes vary very rapidly, too. This would result in a need to adjust the frequency too often. A satisfactory solution from the point of view of the frequency stability seems to be a set of dividers based on phase locked systems. However, the best way is to adjust the frequency automatically. One possible solution is described in section 4.3.

A more effective method of increasing the precision of a measurement can be achieved by using a ratio measurement obtained by computational means. This method enables implicit compensation for atmospheric influences (except air circulation), without the need to measure atmospheric parameters. In comparison with the basic method, this method requires at least one more transducer, placed in a known position. In this way, one more measuring channel is added to the equipment. This (reference) measuring channel can, according to the choice of transducer configuration, carry out the measurement either simultaneously, or sequentially, with the main measuring channel. This method is explained in more detail in section 4.3, for the case of measurement of planar position coordinates.

The measurement shown in Fig. 4.3 is described as a single-shot measurement. During the measurement, which lasts of the order of ms, (e.g. for $x = 1m$, $t_x = 2.9$ ms) a stationary (fixed) position of the receiver is assumed. In this case, the precision can be increased by repetition of the measurement, and calculation of the average value. Suppose that the receiver is moving during the measurement time, the measured value then corresponds to its displacement at the instant of incidence of the ultrasonic wave on the active receiver surface.

On monitoring the trajectory of a moving object, it is recommendable to repeat the measurement in as close a sequence as possible so that the trajectory of the motion can be pursued and reconstructed with satisfactory precision. This holds especially in the case of rapid motion and the requirement for a high-precision measurement. The measurement resembles a sampling with the only difference being that the sampled value is position-dependent, even if the instants of the starting pulses are equally time spaced.

The pulse detection can be modified to give a Doppler detection of the velocity of motion, v, if the pulse generator produces a so-called radiopulse, i.e. a short series of harmonic signals with a given frequency f (burst). The frequency shift, Δf, of the output signal, V_a, is evaluated by an electronic circuit. The relation between the velocity of motion and the frequency shift is expressed by the Doppler equation

$$v = \frac{c}{2f}\Delta f \qquad (4.10).$$

Usually Doppler detection is used for the detection of a reflected wave from the moving object. The receiver is placed close to the transmitter, or is identical, i.e. reciprocal.

Another important improvement in measurement is obtained using a pulse with frequency modulation. For this, the abbreviation CWFM (continuous wave, frequency-modulated) is used. A linearly frequency-modulated ultrasonic pulse is transmitted instead of a simple pulse. The reflected signal is then compared with the original pulse, and the time lag relative to the original pulse is evaluated. The evaluation can be carried out either by simple subtraction of both frequencies, or, in a slightly more complex way, by multiplication of both signals and their subsequent filtering and evaluation [4.5], [4.6].

The time development of the frequency modulated pulse can be triangle-shaped or sawtooth-shaped (Fig. 4.4a). Both the transmitter and the receiver should have a wide frequency range in order to transfer the whole frequency band necessary. For example, a capacitor transducer with a thin layer on a stiff metallic electrode fits well into this scheme [4.5]. In this way, a considerably higher energy can be emitted, and thus the ambient interference is significantly rejected too. The interference can be further efficiently rejected by using filters or even by spectral analysis. This is the principal reason for introducing frequency modulation.

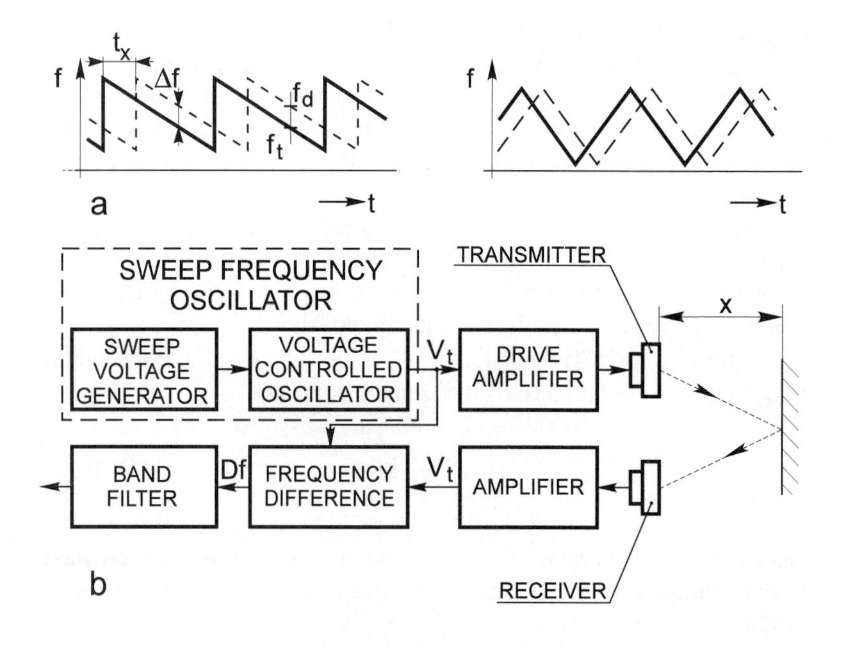

Fig. 4.4 ***Frequency modulated pulse measurement of distance***

The pulse method finds wide application in systems for focusing and deflecting ultrasonic beams (scanning). In medicine, for example, systems for imaging (scanning) of internal body organs by probes with an integrated matrix of transducers are widely used (sonography, echocardiography, etc).

By suitable time-distribution control of the excitation of individual transducers, an ultrasonic beam can be brought to a focus, its focal length can be varied, and deflected in a purely electronic manner. This is illustrated in the simplified diagram in Fig. 4.5. To the left of the transducer matrix, a time shifted series of pulses exciting individual elementary transducers is depicted, and to the right, the time-distribution of the ultrasonic pulse vectors from separate transducers is indicated by arrows, and to the far right, the distribution of the intensity of an acoustic field at the focal distance is illustrated. This resembles visible optics.

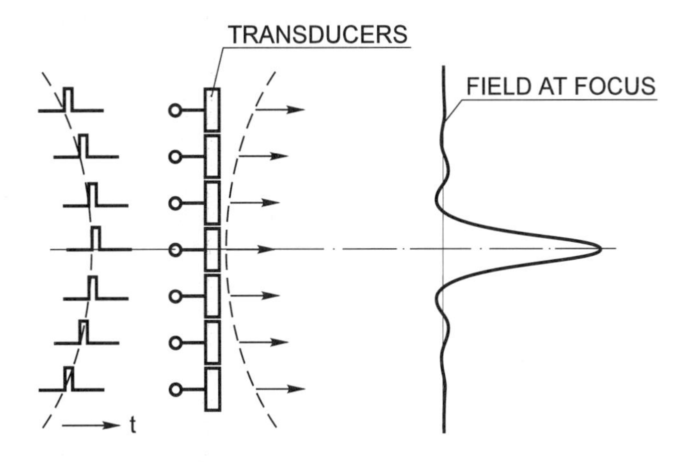

Fig. 4.5 ***A focused ultrasonic wave field***

This branch of application is specialized, and very demanding. It is also extensive, and it is beyond the scope of this book. Therefore, it has been briefly mentioned here only. However, it should be noted that similar systems are used in robotics too. They operate as a kind of sound camera. In contrast to optical cameras, ultrasonic imaging provides one more item of information, namely information about the distance of the object being tracked. It is used for identifying objects, determining their directional and positional orientation, etc. These cameras are also used as an aid for the blind. They provide them with ultrasonic sight [4.5].

When using matrix transducers in regions with a high level of interference (e.g. in factory halls), it is usually impossible to use a very high density of transducers, as in medical applications, for instance. A higher level of emitted energy is used in this case, and therefore the size of the transducers is bigger, which results in a smaller number of them. The processing and evaluating circuits are usually more complex as well.

4.2 Distance and level measurement

Distance measurements belong to the most rewarding and the oldest applications of ultrasonic techniques. In the Munich Technical Museum, an echo sounder from the days before World War 2 may be seen. This equipment represents the first ultrasonic distance meter, either for sea bed depth measurements (an echo sounder*), or for the detection of submarines and measurement of their depth (an underwater echo ranging device; a sonar**). The oldest of all is the equipment designed by Richardson in 1912, after the destruction of the Titanic. This equipment would have been used for recognizing icebergs at large distances, which in those times represented a serious danger to ships.

Besides this widespread and frequent application, at present various room dimension gauges and building-site distance meters are widely employed. A further commonly used application is in level measurement.

4.2.1 Ultrasonic echo sounders

As already mentioned, measurement of the depth of a sea bed or a river bed represents the oldest application of ultrasonic techniques. Figure 4.6 illustrates the principle of operation of an echo sounder. An ultrasonic pulse is emitted in a downward direction from the emitter. After reflection, the echo is detected and processed in an electronic block, and recorded on an X-Y recorder as a time interval that is proportional to the measured distance, or it is depicted on the screen of a CRT, usually in

---

* sometimes called echolot, as an abbreviation from echolocator

** sound navigation and ranging

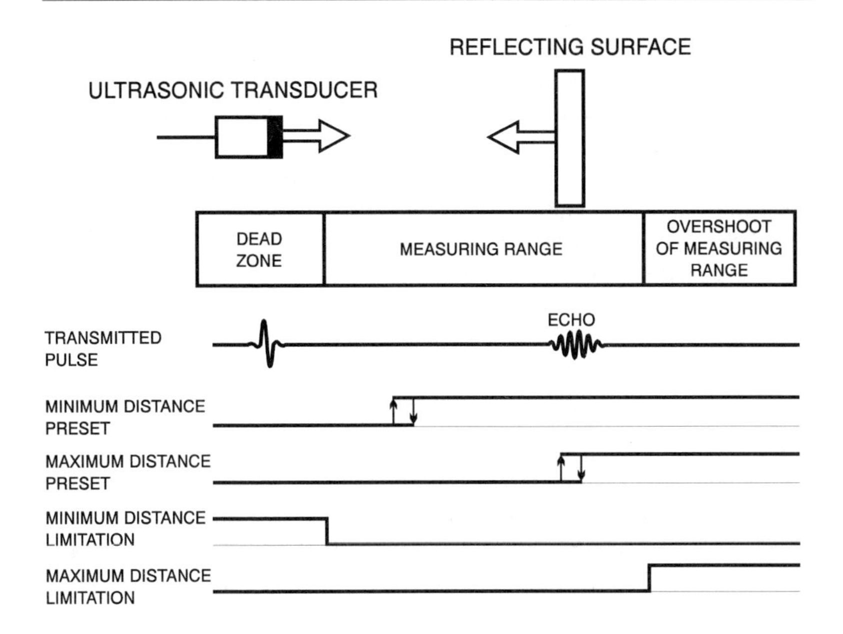

Fig. 4.6 ***Illustration of the principle of an echo sounder***

polar coordinates, in order to obtain a sufficiently long, circumferential, time-coordinate.

Modern equipment evaluates the distance in a straightforward way, showing length units on the display, or the reading is stored directly into computer memory and drawn on a plotter as a chart, showing the contour lines of the sea bed, or the river bed. Ultrasonic echo sounders have a service range from 0-50 m up to 0-10 km, according to their design and specification. The power of an ultrasonic pulse ranges in the interval between 5 W to 50 kW. The working frequency is in the range from 10 kHz to 500 kHz. The instruments are usually equipped with an adjustable length scale in order to compensate for fluctuations of temperature, and propagation velocity of ultrasound, due to fluctuations in the water composition (sea/river water). Ultrasonic echo sounders are widely used thanks to their useful characteristics. Lloyd's shipping register requires that ships sailing beyond a limited distance off the coast have their echo sounders as part of their standard equipment.

The probes of ultrasonic echo sounders have, as a rule, a rugged construction. At the very beginning, magnetostriction transducers were used most often for this purpose. Until recently, they proved successful, especially with high-power probes of the order of 10 kW. At present, piezoelectric transducers are now used for the same values of power. They are easier to manufacture and they are more efficient too. The design is usually of the sandwich type, with a unit which is able to process a single pulse power of 10 kW. To obtain higher values of power, several individual transducers are arranged into a system.

4.2.2 Ultrasonic length meters

Due to the recent achievement of very good results regarding emission of ultrasound in air, a whole series of interesting new applications have appeared. One of these is distance measurement in the building industry. Pocket instruments with digital signal processing enable distance measurement with an accuracy of 1 cm, over a range 0-20 m. The system is based on the principle of measuring the time of propagation of an ultrasonic pulse to an obstacle and back to the receiver. The built-in microprocessor makes it possible to convert the measured time into a length in cm, to store a group of numbers and to execute a series of operations with them, e.g. to calculate the cubic capacity of a room, to compensate for the effect of the propagation velocity with respect to temperature, etc. The ultrasonic transducers of these systems operate with bending waves. Separate emitting and receiving transducers are used in this case in order to achieve a minimum dead zone.

To extend the service range, the probes are equipped with a parabolic reflector. These instruments are usually battery-operated.

4.2.3 Ultrasonic level meters

Besides other methods employing ultrasound, measurement of liquid levels can also be carried out on the basis of measuring the propagation time of an ultrasonic wave to the liquid level and, after reflection, back to the receiver. The propagation time of the wave is proportional to the level distance,

$$l = t.c/2 \qquad (4.11)$$

where l – distance level-transducer
t – propagation time of an ultrasonic wave to the level and back to the receiver
c – velocity of propagation of the ultrasonic wave.

Devices constructed that utilize this method enable continuous measurements on liquids, or loose materials. In principle, a periodic transmission of ultrasonic pulses is used in this case and the time delay between the emitted pulses and the reflected ones is measured.

According to the positioning of the transducers, instruments for measuring level can be divided into the three following groups:

meters using wave motion which propagates in

1) liquid
2) a sound guide (an acoustic horn)
3) gas.

In the first group, the physical phenomenon that the transducers radiate well into the liquid is used, which enables low-power operation, high working frequencies (of the order of MHz) and relatively low amplification at the receiver. A drawback of these level-meters is that they cannot be used in cases when:

a) there are mud discharges, or drains, and other mechanisms at the bottom of the storage tanks
b) the wave motion is heavily attenuated by the liquid
c) the liquid contains solid particles.

The group of ultrasonic level-meters which use a fixed, rigid sound-guide (sound-horn) has similar advantages to the first group, and in addition they enable level measurement even in the cases when there are drains at the bottom of a storage tank, or when there are solid particles dispersed in the liquid. A particular drawback, however, is the inability to measure levels of liquids with a viscosity higher than about 100×10^{-5} Pa.s, when the inaccuracy of measurement begins to increase rapidly.

Ultrasonic level-meters which use the propagation of ultrasonic waves in gas do not have the handicaps of the first group, and they allow one to monitor levels of loose and pulpy materials as well. On the other hand, the attenuation of ultrasonic waves in a gaseous medium demands considerable transducer power, and working frequencies at the bottom end of the ultrasonic frequency band, even down to audible frequencies.

The relatively high dependence of ultrasound propagation in a gaseous medium on temperature reduces the precision of measurement.

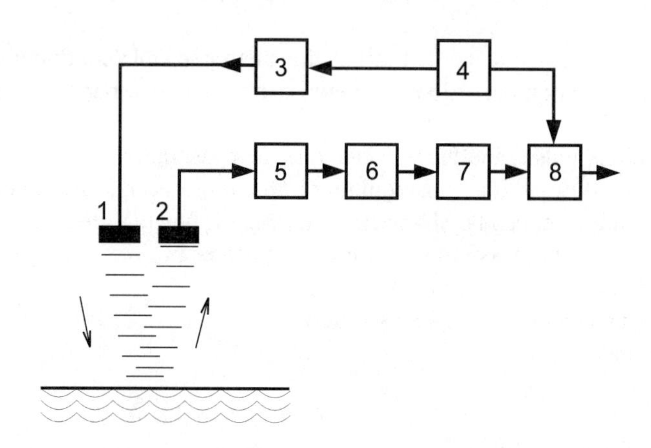

Fig. 4.7 ***Illustration of the principle of an ultrasonic level gauge: 1 - emitting transducer, 2 - receiving transducer, 3 - excitatory generator, 4 - pulse source, 5 - amplifier, 6 - detector, 7 - shaper, 8 - bistable circuit***

The principle of an ultrasonic level gauge is illustrated in Fig. 4.7. The excitatory generator 3 of the emitting transducer 1 is controlled by pulses from the pulse source 4. The ultrasonic pulse propagates through liquid (or through gas, if the transducers are placed above the level) to the level, and, after reflection, it returns back to the receiving transducer. The voltage pulse from the receiving transducer is amplified in the amplifier 5, and after detection in the detector 6 it is led to the pulse shaper 7. The bistable circuit 8 is controlled by this shaped pulse. The initial control pulse from the source triggers the bistable clamp, and the pulse detected by the receiving transducer returns the bistable to its initial state. The duration of a pulse generated by the bistable circuit is proportional to the time of the pulse propagation in liquid, or in gas, and is thus proportional to the transducer-level distance.

The transfer of the wave motion into the medium, its reflection and its reception are affected by the following factors:

- the transducer-medium coupling
- the transfer of ultrasound through the medium above the liquid level
- the condition of the reflecting surface.

Upon transfer of an ultrasonic wave from one medium into another, the ratio of acoustic wave resistances plays a critical role. The wave resistance is expressed as a product of the density of the medium and the velocity of the wave, $Z = \rho C$, and it represents a characteristic feature of a particular material. The wave resistance of air is 430 $kgm^{-2}s^{-1}$, and that of water from 1.49×10^6 up to 10^7 $kgm^{-2}s^{-1}$. At the interface of two media with different acoustic impedances, Z_1 and Z_2, part of the wave motion is reflected back and another part transfers into the second medium.

The coefficient of reflection, R, i.e. the ratio of the intensity of the reflected wave over the total intensity of the primary wave, is given by

$$R = \left(\frac{Z_2 - Z_1}{Z_2 + Z_1}\right)^2 \qquad (4.12).$$

The coefficient of transmission, D, i.e. the ratio of the intensity of the transmitted wave to the total intensity of the incident wave, is:

$$D = 4\,\frac{Z_1 Z_2}{(Z_1 + Z_2)^2} \qquad (4.13).$$

Thus, for a wave to be transferred from one medium into another efficiently, it is necessary to choose the physical properties of a medium so that the expression for R is as small, and the expression for D as high, as possible. This case occurs when the values of the acoustic wave impedances, Z, of both media are close to each other.

In the case of the transfer from a transducer into a gaseous medium, however, differences of wave resistances up to 10^6 can be met. Therefore, matching elements must be used, so that the input impedance approaches the acoustic impedance of the medium.

During propagation, an ultrasonic wave is absorbed by the gaseous medium, and its acoustic energy is converted to heat.

The intensity, I_x, of the wave motion at a distance x from the source (with intensity I_o, at a distance $x = 0$) is:

$$I_x = I_o e^{-2f^2\alpha' x} \tag{4.14}$$

where f - frequency
$\alpha = f^2\alpha'$ - absorption coefficient.

For example, at a frequency of 30 kHz and a distance of 30 m, the intensity will drop to one half of the initial intensity. The absorption coefficient depends on the air temperature and humidity. In Fig. 4.8, the dependence of the absorption coefficient on frequency, at different values of the relative humidity, and at a temperature of 20 °C, is shown. For example, at a frequency of 25 kHz, a change of the relative humidity from 20 % to 90 % will cause a change of the absorption coefficient by a factor of three.

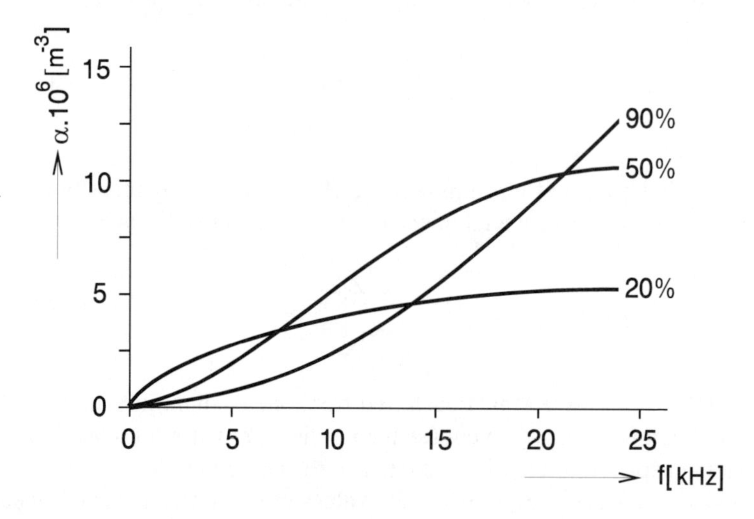

Fig. 4.8 ***Dependence of absorption coefficient on frequency***

Since the level of pulpy, sticky, and highly viscous liquids is flat, and the ratio of the impedances of air and a typical liquid is about 10^5, the reflection coefficient is $R \cong 100$ %. Due to this total reflection, practically all the acoustic energy is reflected back to the acoustic emitter. Thus for this application a lower level of emitted energy than for loose materials is sufficient.

In a gas the velocity of propagation of an ultrasonic wave depends on the temperature and the density. Thus

$$c = \sqrt{\varkappa \frac{p_0}{\rho_0}(1+\gamma)} \qquad (4.15)$$

where p_o - pressure
ρ_o - density
γ - thermal expansion coefficient of gas (approximately 1/273 for all gases)
$\varkappa$ - Poisson's constant.

For air, the following values are known (for 0 °C):
$\rho_o = 1.29\ kg.m^{-3}$
$c = 330\ m.s^{-1}$
$\varkappa = 1.4.$

The mean value of pressure is about 101 kPa (760 Torr). The dependence of density versus temperature is given in Fig. 4.9. It can be approximated by a straight line equation. Then the empirical relation for the dependence of density on the temperature is

$$\rho = A - k\Theta \qquad (4.16)$$

where $A = 0.00129.$

Thus for the velocity of propagation one may write:

$$c = \sqrt{\varkappa p_0 \frac{1+\gamma\Theta}{A-k\Theta}} \qquad (4.17).$$

The dependence c = f(t) for air is shown in Fig. 4.10. A temperature change from 0 °C to 50 °C results in a change of the output value of the level-gauge by 10 %. Such an error would impair the measurement in practice.

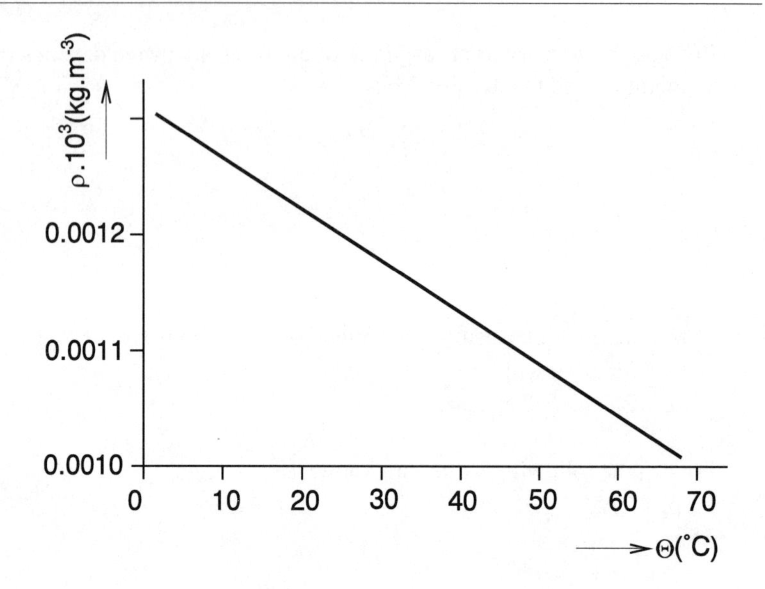

Fig. 4.9 ***Dependence of air density on temperature***

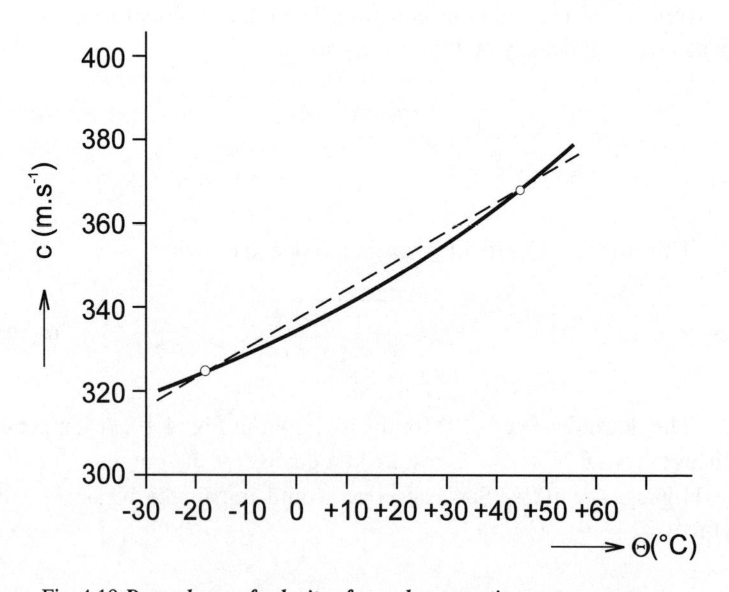

Fig. 4.10 ***Dependence of velocity of sound propagation on temperature***

It can be seen from the dependence of the propagation velocity on temperature (Fig. 4.10) that the curvature of the characteristic is not too strong. Over a given interval, it is possible to approximate it by a straight line.

A maximum deviation of about 0.6 % occurs at a temperature of 20 °C. Hence it follows that this problem can be solved without causing too large an error, by introducing a compensation for the level-gauge output signal based on temperature measurement, for example, using a thermistor.

It follows from the above analysis that the accuracy of a measurement is influenced substantially by temperature fluctuations. This effect can be compensated by measuring the ambient temperature. Such compensation suits analog measuring devices.

In measuring the level of loose materials, the receiving transducer detects a diffusely reflected signal. The resultant wave motion after diffuse reflection is an additive result of interference of waves reflected from elementary, randomly oriented reflecting faces (4) (Fig. 4.11), which create the level surface. At a wavelength comparable with the characteristic size of the average surface roughness, waves diffracted on the edges of the elementary uneven faces also contribute to the resultant signal. As a result, the character of the diffuse reflection depends

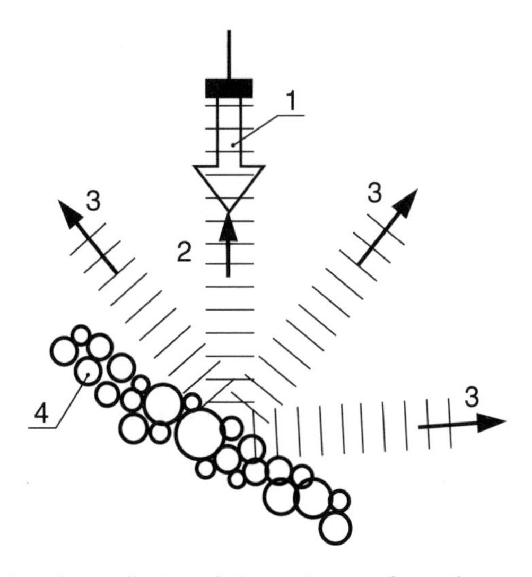

Fig. 4.11 ***Illustration of the reflection of ultrasonic waves from a loose material surface***

considerably on the ratio of the wavelength of the incident wave to the characteristic size of the roughness, as well as the angle of wave incidence. The incident wave motion (1) (Fig. 4.11) is scattered approximately isotropically (3), and only part of it (2) is reflected back to the receiver. From experimental measurements, it follows that due to diffuse reflection the attenuation of an ultrasonic signal can be as high as 25 dB.

For determination of the acoustic pressure of a source, the estimation begins at the level of interference of the ultrasonic signal. In the frequency range up to 20 kHz, the interference level is about 36 dB. The attenuation caused by divergence of the ultrasonic beam is 26 dB, at a divergence angle of ± 3°. Hence the effort is obvious to create as small a transmitter radiating angle as possible. The attenuation caused by absorption over a distance of 40 m is about 30 dB. Due to diffuse reflection at a beam divergence of 3°, the additional attenuation is 25 dB. So the total attenuation over the entire path emitter-loose material-emitter is 81 dB. In addition, to reach a signal-to-noise ratio of 1:1, the emitting transducer must supply a signal with a level of 81 + 36 = 117 dB (considering the base interference level 36 dB). The S/N ratio 1:1 is a limiting value for processing the signal, and if an inevitable surplus of power is assumed, the intensity of the emitted wave should be at least 3 times higher.

Collecting information about the volume of liquid in various storage tanks and containers it is assumed that one has equipment for checking these values sequentially. The first requirement in implementing such a solution is to have a level meter, or at least its probe, suitable for constructing separately inside the container and for carrying out the evaluation in a common, central unit. Since a small error in level measurement causes a relatively high error in the volume measurement, one can understand the user‘s requirement for an accuracy of millimeters. In manual measurement, a precision of the order of cm is usually achieved.

From the following example, an estimate can be made about the requirements on the precision of a measurement. Let the height of a storage tank be 10 m, and its diameter be 4 m. Then the total volume is 125.6 m^3. A level change of 1 mm means a volume change of 12.56 l of liquid. It can easily be shown that if the measurement accuracy of a measurement of 1 mm is required, then the relative precision of the measurement is 0.01 %. It is obvious that only a digital measuring technique can secure such precision.

One of the possibilities for level measurement is to utilize the propagation of an ultrasonic pulse through a liquid. The principle of measurement is based on the emission of periodic ultrasonic pulses towards the liquid level and measuring the time delay between the emitted and the reflected pulses.

One advantage of ultrasonic level gauges is that in contrast to float-type meters, there are no floats to get jammed. These level meters are safer than those using radioactive sources. They do not contain moving mechanical parts, and consequently they are suited to measurements in special surrounding materials.

For ultrasonic liquid level meters, for a particular medium, the most important interfering effect is the temperature dependence of the velocity of wave propagation. This dependence is expressed as

$$c = c_o + b \cdot (\Theta - \Theta_o) \tag{4.18}$$

where c_o - velocity of the ultrasound propagation at the temperature t_o
b - temperature coefficient of velocity, c/t.

In all pure liquids, the velocity coefficient b is negative. In water, the temperature dependence of ultrasound propagation velocity is non-linear, and is given as

$$c_t = c_{max} - 0.0245(\Theta_{max} - \Theta)^2 \tag{4.19}$$

where c_{max} - the maximum velocity of ultrasound in water, $c_{max} = 1557\ ms^{-1}$
Θ_{max} - temperature of water, corresponding to the maximum velocity of ultrasound in water, which occurs at 74 °C.

It follows from Table 2.1 that the velocity of propagation varies in different liquids, and at different temperatures, in practice, from 1000 to 2000 ms^{-1}. Thus compensation is required in this range.

After adjusting the level meter for a particular liquid, a measurement error occurs due to variations in temperature. In the reflection method, where the time of the pulse propagation to the liquid level and back to the emitter is measured, the level distance is given by

$$h = \frac{1}{2} . ct \qquad (4.20)$$

where t - time of propagation of the ultrasonic pulse
c - velocity of ultrasound propagation in the given medium
h - the distance between the liquid level and the transducer.

Regarding the temperature dependence of the propagation velocity, after substituting (4.18) into (4.20), we obtain for the distance h:

$$h = 1/2 . [c_o + b(t - t_o)]t \qquad (4.21).$$

It follows from the above that for ultrasonic wave propagation in fuel oil, for instance, the pulse propagation time is t = 10 ms at a level-transducer distance of 6.76 m, and at a temperature 20 °C. Increasing the temperature to 60 °C, gives rise to an error and at a measured propagation time of t = 10 ms, the result of the level distance measurement is 6.14 m. The error of 0.62 m is a high value. Using the dimensions of a storage tank from the foregoing example, it would represent 7787.2 l of fuel oil.

Since the change in the ultrasound propagation velocity is not always linear, this must be considered in choosing the measuring method. The most universal method is based on comparing the time of propagation or the frequency of a pulse circulating through a closed loop of two channels, namely the measuring channel (transducer to level) and the reference channel (transducer to reflector). In the first case, the distance of the ultrasound propagation is unknown and variable, in the second case, the distance is known and constant. For both channels, the following relations can be written:

$$l_x = 1/2.[c_o + b(\Theta - \Theta_o)]t_x \qquad (4.22)$$

$$l_l = 1/2.[c_o + b(\Theta - \Theta_o)]t_l \qquad (4.23).$$

By comparing these relations, the level distance, l_x, is obtained from the expression

$$l_x = l_l \frac{t_x}{t_l} \qquad (4.24)$$

where l_l is a constant and the time intervals t_x and t_l correspond to the propagation times of an ultrasonic wave in the measuring and reference channels respectively. Thus we obtain the level distance value, independent of the velocity of propagation of the ultrasound in a particular medium.

4.3 Position measurement

Mechanical quantities like distance, position, movement, thickness, velocity, etc. belong to a group of very frequently measured physical quantities. This is why the measurement techniques for these quantities are also the most elaborate ones. This is especially the case for length measurements and straight-line position measurements. These one-dimensional measurements are most frequently taken. Measurements of other physical quantities are indirectly converted to these types. Multi-dimensional measurements are, as a rule, more complex and less accurate. In practice, multidimensional measurements are carried out most often as separate measurements of individual dimensions, or coordinates. In such an arrangement, the pick-ups or sensing probes are mechanically connected to one another. This can be performed relatively easily in a plane. In 3 dimensions, this method is considerably more demanding, though.

Compared with other methods, the ultrasonic ones allow one to design and to realize relatively simple instruments for position measurement, in 2D as well as in 3D. This is due to the aforementioned feature of ultrasonic waves that they propagate slower by about 6 orders of magnitude than electromagnetic waves. In addition, sources of spherical and cylindrical ultrasonic waves can be produced easily enough by modern transducer technology. Several kinds of ultrasonic transducers are reciprocal, so that they can be used both as receivers and as transmitters at the same time (see Chapter 3). These beneficial properties have caused a rapid development of 2D and 3D ultrasonic measurements.

In general, for measurement of position and dimensions, wave-based methods are the most suitable ones. Both ultrasonic and the electromagnetic waves have their advantages and drawbacks. In electromagnetic measurements, a wider frequency band is used. Both types of measurements are contactless. A drawback of electromagnetic measurements compared to ultrasonic ones is that due to the high propagation velocity: they do not allow application of small-distance pulse methods and there are no suitable sources of spherical waves. Therefore, in the cases of position measurements, a time shift of a stationary electromagnetic field (or, alternately, an electrostatic field too), is applied. For example, we compare two methods of measuring plane coordinates.

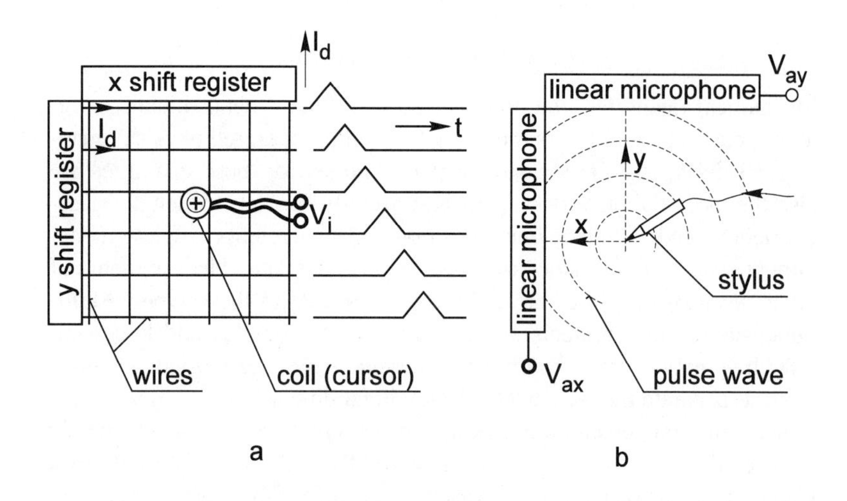

Fig. 4.12 ***A schematic illustration of the principles of measurement of point coordinates in a plane***

In Fig. 4.12, the most frequently used methods, exploiting the two types of wave motion (electromagnetic and ultrasonic), are illustrated in a simplified way. The measurement utilizing the electrostatic field is very similar to the electromagnetic one. The shift of the electrostatic field is achieved by sequential switching of current into two mutually perpendicular systems of sets of parallel conductors (Fig. 4.12a). The coordinate measurement cannot be carried out by simultaneous switching-on of the current into the system of perpendicular conductors, therefore the coordinate measurement must be divided into two sequential phases (step by step measurement). First, one coordinate is measured, then the other one. In the adjacent figure, a time diagram is plotted for the sequential switching of current and its time development in separate conductors, for measurement of the y-coordinate. When the magnetic field, which is constantly varying in time, gets close to the pick-up coil (a cursor), which is placed above the conductor system (termed a pad plate), a voltage is induced in the pick-up coil. From the voltage time development, a coordinate can be evaluated because the displacement as well as the time shift rate are controlled electronically.

In a case of ultrasonic sensing (Fig. 4.12b), a spherical or a cylindrical ultrasonic pulse wave is excited by a transducer, placed above the detected point (e.g. by a small spark between an electrode and the tip of a ball point pen). When the front face of the pulse touches the active surface of a linear microphone, a voltage is induced in the transducer. The instant of arrival of the voltage pulse corresponds to the shortest (i.e. perpendicular) distance between the detected point and the linear microphone. The time interval, t_x, from the instant of spark occurence until the arrival of the front face of the ultrasonic pulse wave is evaluated and converted to the coordinate x, according to equation (4.9). The y coordinate is evaluated in the same manner.

By comparing the two methods described above, some advantages of the ultrasonic method of coordinate measurement can be derived. When using an electromagnetic wave, positioning a set of conductors below the detected surface (i.e. below the pad plate) is necessary. In ultrasonic sensing, transducers can be placed at the outer borders of the examined surface (or, if necessary, at only one of them), so that the examined surface, and the pattern to be digitized, is freely accessible. Thus, in the first case, the thickness and the type of material on which the pattern is deposited influences the precision of the measurement. The pattern cannot exceed the dimensions of the conductor set. In the second case, the pattern size may exceed the position of the sensing probes, especially if point microphones are used.

4.3.1 Plane coordinate pick-ups

The method described above of measuring coordinates by using linear microphones has its advantages in terms of simplicity and ease of use. However, with increasing size of the measured object, problems with ensuring the flatness of the active pick-up surface increase too. Usually this surface consists of a thin metallized insulator foil (this may be gold-plated or silver-plated styroflex, mylar, etc.) deposited on a stiff basic metallic electrode.

The influence of fluctuations of the medium parameters on the precision of a measurement can be reduced in several ways. The simplest methods were mentioned in section 4.1. The methods which compensate several parameters at a time are more efficient. One of the possibilities is to control the frequency of the time mark (clock) generator. As the time stability of frequency-controlled oscillators is rather low, it is recommendable to regulate the frequency by using an automatic control circuit and include the possibility of fine tuning. One such way is shown in Fig. 4.13.

A pick-up, illustrated in Fig. 4.3, serves as a central part of the control circuit. An auxiliary pick-up is placed close to the examined surface. Its basic electronic circuit has already been described. The transducers are stationary and they are positioned at a known distance. After the reference measurement is over, a number N_r, proportional to the preset distance x_r, appears at the output of the electronic circuit. The control circuit consists of a digital comparator, an analog-to-digital converter, and a reversible counter [4.8].

To one input of the comparator, a number, N_R, is conveyed, which corresponds to the preset distance of the transducers. The measured number, N_r, is passed to another input. The comparator generates a pulse train and its count, N_d, is proportional to the difference $N_d = N_R - N_r$. These pulses are counted by a reversible counter which changes the previous state of the analog-to-digital converter. A result of this regulation process is a tuning in (adjustment) of the frequency of the clock pulse generator. This frequency is used in both circuits for coordinate measurement, for conversion of the time measuring pulse to a number.

The adjustment method described above could, in principle, be performed manually, which would simplify the circuit. However, this way would be more demanding of the operational staff. It would be sufficient, for example, to place the cursor at a known position at certain intervals, and the frequency could be adjusted manually, until the output reading showed a known (preset) value.

From the point of view of compensation of atmospheric influences, such comparative measurements would be desirable which would enable one to compensate for all climatic effects in the measured space, including air flow. This is enabled by the introduction of reference transducers [4.9]. In Fig. 4.14, a possibility is illustrated, which uses a reference

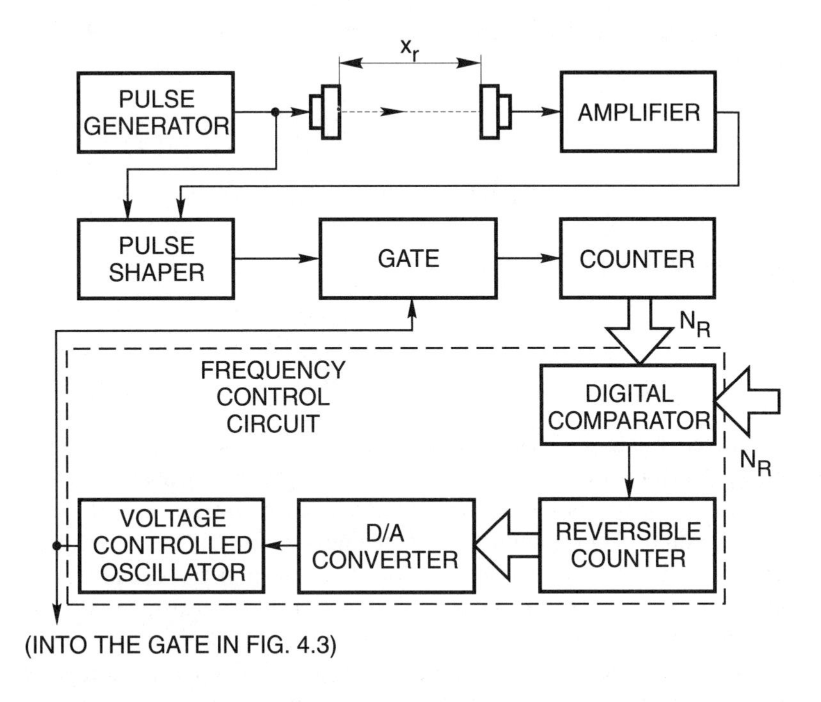

Fig. 4.13 ***A circuit for automatic adjustment of the frequency of the clock pulse generator***

transmitter. It is an extension of the arrangement shown in Fig. 4.14b. For simplicity, the electronic circuit has been drawn here merely as a single block (for more details, see Fig. 4.13). It is assumed that miniature spark gaps are used as sources of ultrasonic wave pulses. Moreover, a microprocessor is introduced, which is necessary for performing the arithmetic in the coordinate calculations. The measurement is carried out in a two-step fashion. In the first phase, a measurement with a reference transducer is performed at a point with the coordinates x_r, y_r; in the second phase, a transducer T_m is used at a point with the known coordinates x_m, y_m. In both phases, the same linear transducers together with their appropriate electronic circuits are used. In the first phase, the numbers N_{xr}, N_{yr} are obtained and stored in the microprocessor memory. In the second phase, the numbers N_{xm}, N_{ym} are measured and they are transferred to the memory as well. Then calculation of the coordinates follows. Neglecting the discretization error in the expression (4.9), one may write

$$x_m = k_t N_{xm}\ ; \quad y_m = k_t N_{ym}$$

and

$$x_r = k_t N_{xr}\ ; \quad y_r = k_t N_{yr}.$$

By dividing the equations, we obtain

$$x_m = \frac{N_{xm}}{N_{xr}} x_r\ ; \quad y_m = \frac{N_{ym}}{N_{yr}} y_r \qquad (4.25).$$

The coordinates of the measured point can be calculated after introducing the ratio of numbers, measured during the two phases, at the known coordinates of the reference point. Thus the influence of fluctuations of atmospheric parameters, including air flow, is fully compensated. This is rather important because air circulation can occur also in laboratory conditions and reduce the precision of a measurement (e.g. draughts, fans, air conditioning, breathing, etc).

However, placing the reference transducer at the opposite corner of the detected area is not reasonable from the practical point of view, because it can get in the way while a measurement is carried out.

This obstruction can be avoided by the arrangement shown in Fig. 4.14b. A reference transmitter, T_r, is placed in the opposite corner of the operational area. There it does not get in the way; however, two more recievers, R_{xr}, and R_{yr}, are necessary as well as the appropriate electronic evaluation circuits. On evaluation of the coordinates, the influence of air parameters is compensated, except for air flow. The transducers, however, can be built in together with the linear microphones, in their common covers. Thus removal of the emitting transducer from the operational area makes the instrument more complicated, lowers its measurement precision, but allows easier access to the measured space.

For completeness, let us mention briefly another way of introducing ratio measurement, by doubling the number of linear microphones (Fig. 4.15). As in the foregoing case, relationships can be derived for the coordinate calculation:

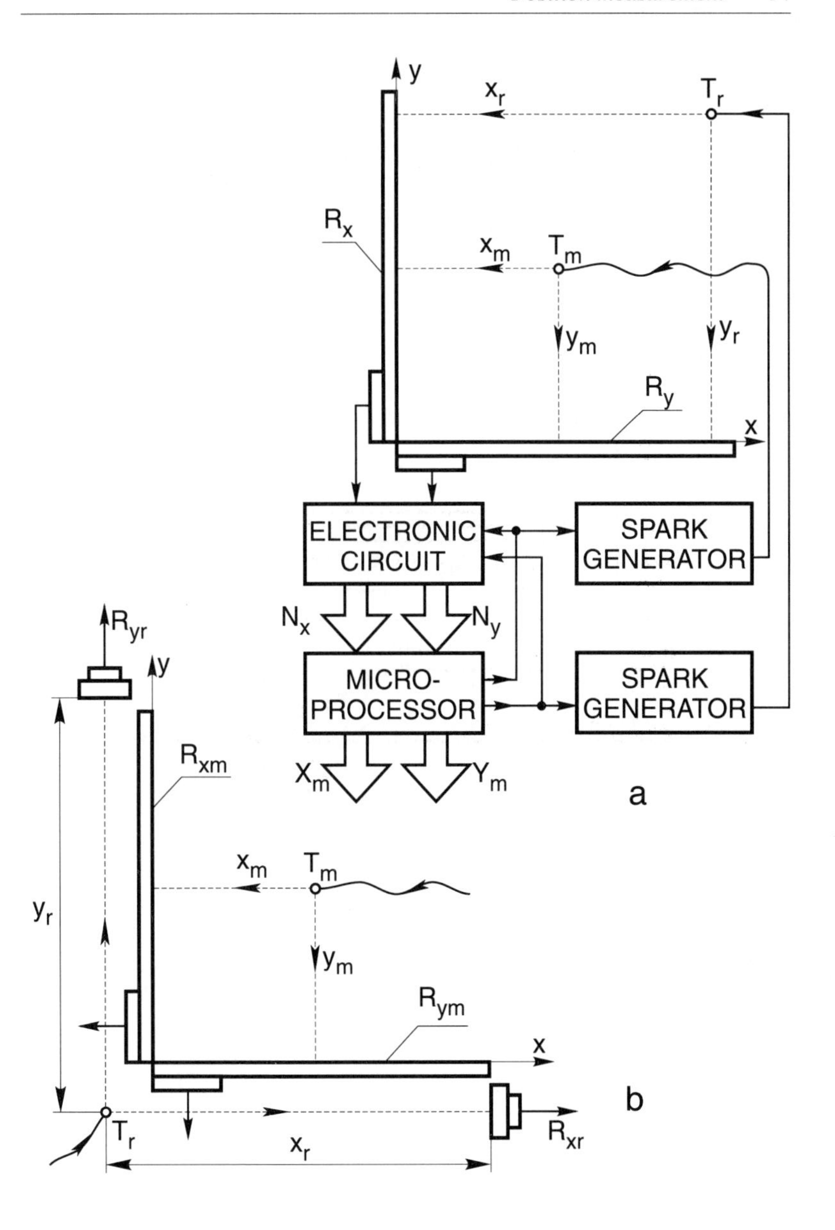

Fig. 4.14 ***Arrangement of the coordinate pick-ups with reference transducers***

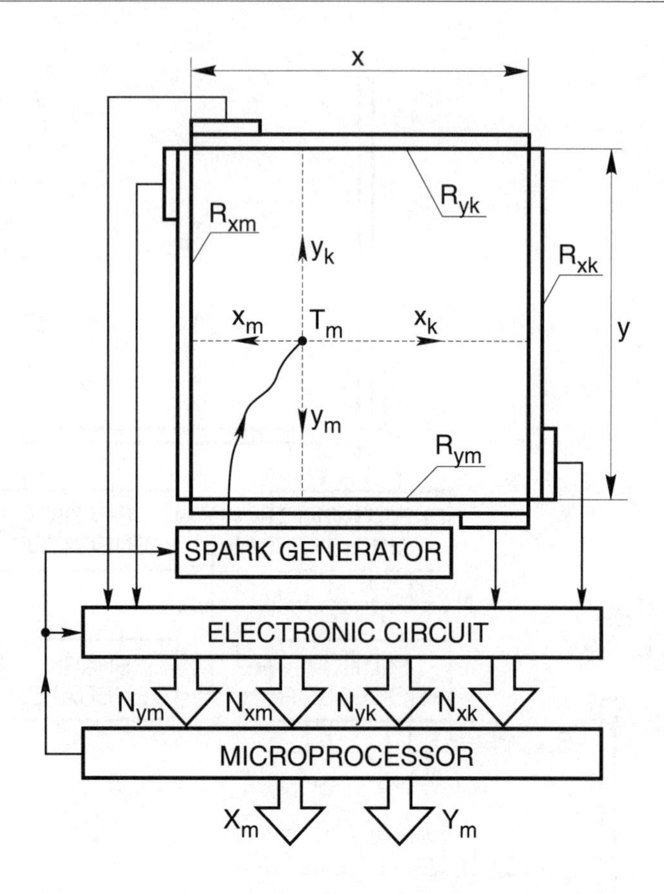

Fig. 4.15 ***A coordinate pick-up with a system of doubled linear microphones***

$$x_m = \frac{N_{xm}}{N_{xm} + N_{xk}} X$$

$$y_m = \frac{N_{ym}}{N_{ym} + N_{yk}} Y \qquad (4.26).$$

It can be concluded directly from the diagram of the system that placing the linear microphones on the perimeter of the detected area will complicate the job of coordinate taking. As in the previous case, the pick-up is equipped with four complete measuring channels including four linear microphones (a difference from the foregoing case). The need for a single transmitter can be considered as an advantage of such a pick-up combination as well as only having a single-phase measurement. The influence of the air circulation is not compensated in this case, though.

In order to avoid drawbacks which come from the need to use linear microphones, various configurations of transducers have been invented and patented. These inventions simplify to a greater or lesser extent the design of the coordinate pick-ups and increase the precision by introducing various calibrations, corrections, compensations, etc. (for instance, as graphical inputs for computers, also called digitizers). Their outputs are mostly matched for direct connection to computers.

In Fig. 4.16a, the principle of coordinate measurement with two quasi-point microphones is indicated [4.7]. From the measured distances d_x, d_y, and using the distance, D_R, between the receivers R_x, R_y, the coordinates can be calculated. An advantage of such an arrangement is that a supporting pad is no longer necessary for precise in-plane fastening of extended microphones. It is sufficient to secure the microphones in a mutually constant position. This can also be done by fastening them to the common frame housing the electronic circuitry.

If Fig. 4.16a is redrawn as shown in Fig. 4.16b, an analogy can be found with Fig. 4.12b. Instead of two transducers (linear microphones), two pointmicrophones are placed here, in two virtual coordinate axes, at distances D_x, D_y from the coordinate origin. Neglecting the discretization error, one may write

$$d_x = k_t N_x \,; \qquad d_y = k_t N_y.$$

Using Fig. 4.16b, a system of equations may be written

$$(D_x - x_m)^2 + y_m^{\;2} = k_t^{\,2} N_x^{\;2}$$

$$(D_y - y_m)^2 + x_m^{\;2} = k_t^{\,2} N_y^{\;2}$$

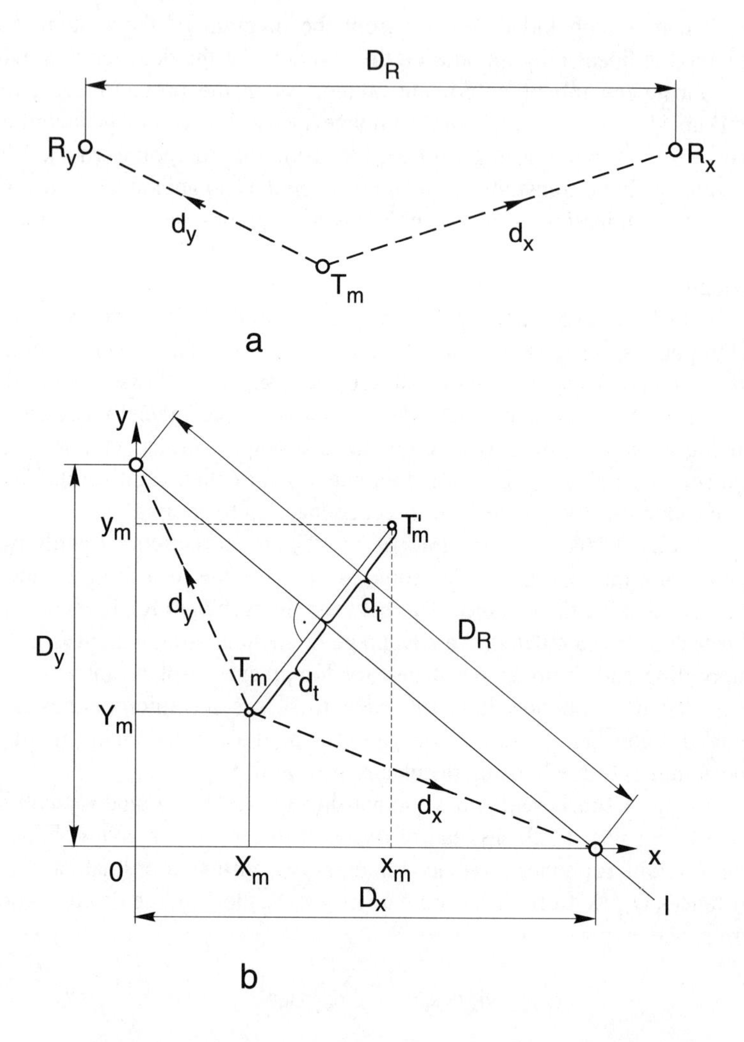

Fig. 4.16 ***Coordinate measurement using two point microphones***

By solving these we obtain the formulae for calculation of the coordinates (for simplicity, $D_x = D_y = D_{xy}$ is chosen here)

$$x_m = \frac{1}{4D_{xy}}\left[2D_{xy}^2 - k_t^2\left(N_x^2 - N_y^2\right) \pm \sqrt{D}\right]$$

$$y_m = \frac{1}{4D_{xy}}\left[2D_{xy}^2 - k_t^2\left(N_y^2 - N_x^2\right) \pm \sqrt{D}\right] \qquad (4.27)$$

where $D = 4k_t^2 D_{xy}^2(N_x^2 + N_y^2) - k_t^4(N_x^2 - N_y^2)^2 - 4D_{xy}^4$.

After carrying out the calculation, coordinates of two points are obtained - a real one at the point T_m, and its mirror (virtual) image along the straight line, l, passing through the transducers R_x and R_y. Unambiguous coordinates are obtained only if the measurement is limited to a half-plane.

It is obvious that we have obtained more complex formulae than in the previous cases. However, they do not represent a significant obstacle to modern computing techniques. More of an obstacle is the fact that the distribution of the accuracy is not uniform because of the non-linear relationships. The same holds true for the influence of fluctuations of atmospheric parameters.

4.3.2 *Pick-ups of spatial coordinates*

The main advantages of ultrasonic sensing, in contrast to other ways of coordinate measurement, are fully revealed when measuring spatial (3-D) coordinates. The relatively easy ways of producing transducers for generating spherical ultrasonic waves is exploited here. This allows direct and simultaneous measurement of all three coordinates. Whilst in planar coordinate measurement it was possible to generate a travelling magnetic or electrostatic field by a system of mutually perpendicular sets of conductors, a spatial analogue of this is not possible.

Using planar microphones (see Fig. 4.17a) means an extension of plane coordinate sensing to spatial sensing. The values of individual coordinates are in this case determined similarly to the case of planar coordinate measurement, by linear microphones (Fig. 4.17b).

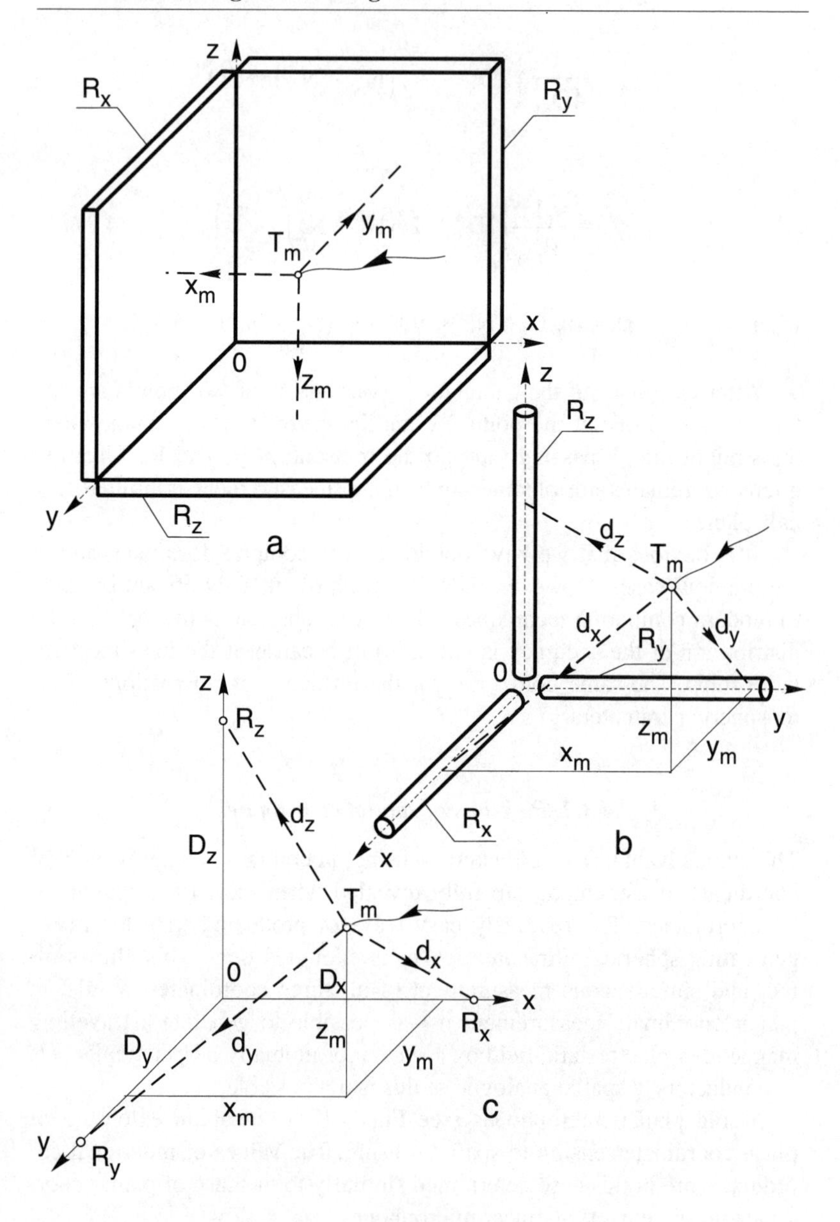

Fig. 4.17 ***Methods of measuring spatial coordinates***

$$x_m = c/f.N_x = k_tN_x\,; \quad y_m = k_tN_y\,; \quad z_m = k_tN_z \qquad (4.28).$$

Apart from the effect of other influences, the measurement error also depends on the velocity of sound propagation, c. The frequency of the clock pulses, f, can be maintained to a high level of accuracy. To reduce the errors, the same techniques can, in principle, be used as described in section 4.3.1. However, from the practical point of view, some of them are unusable. Measuring spatial coordinates with planar microphones is suitable only in relatively small measurement volumes. For larger volumes, the design becomes clumsy and the sensitivity of the microphones decreases inversely with their size. In the ideal (theoretical) case, they should react to the point incidence of the front face of an incident ultrasonic wave. Actually, the reality is a miniature (quasi-point) area. However, even this miniature surface area represents a tiny, insignificant part compared with the area of a planar microphone and, in this regard, also a negligible change of capacitance. From this point of view, the design of the sensing part using cylindrical linear microphones as shown in Fig. 4.17b seems to be more suitable. Although their sensitive area is smaller than in the case of flat microphones, they are favoured because of the high demands being placed on the detection of the front face of an ultrasonic wave pulse. For the calculation of coordinates, the following relations are presented [ref.4.10]:

$$\begin{aligned} x_m &= \frac{1}{\sqrt{2}}\sqrt{(k_tN_y + r)^2 + (k_tN_z + r)^2 - (k_tN_x + r)^2} \\ y_m &= \frac{1}{\sqrt{2}}\sqrt{(k_tN_x + r)^2 + (k_tN_z + r)^2 - (k_tN_y + r)^2} \\ z_m &= \frac{1}{\sqrt{2}}\sqrt{(k_tN_x + r)^2 + (k_tN_y + r)^2 - (k_tN_z + r)^2} \end{aligned} \qquad (4.29)$$

where r – radius of the cylindrical linear microphones

N_x, N_y, N_z – values obtained by measuring the perpendicular distances d_x, d_y, d_z.

A further simplification of the sensor system is illustrated in Fig. 4.17c. Miniature (quasi-point) microphones R_x, R_y, R_z are used there and they are positioned on the virtual coordinate axes, at known distances D_x, D_y, D_z from the origin of the coordinate system. The microphones should have a wide directional pattern and a high sensitivity. Microphones as well as transmitters can be spherically sensitive. Although their manufacture is slightly more demanding and the evaluation formulae are more complex, their chief advantage lies in their omnidirectional characteristics and in a more precise localization of the spherical wave as well as in its detection. On the assumption of point transducers, and with the simplification $D_x = D_y = D_z = D_{xyz}$, relations similar to the formulae (4.27) are obtained, but still more complex.

When using spherical transducers, the calculation gets more complicated. Indeed, using subminiature (quasi-point) microphones gives rise to errors; however, the error is negligible at distances considerably larger than the active surface area of the microphones. With the advances in the technological production of subminiature microphones, the number of types of coordinate sensors has increased, based on their utilization.

The design of the sensors described so far does not enable direct compensation of the influence of environmental conditions. In Chapter 6, one possible solution of this problem is described in more detail, enabling compensation of all atmospheric effects including air flow, by using an auxiliary microphone added to the system, according to Fig. 4.17c.

Over a period of time, many sensor set-ups of various spatial coordinates have been invented which utilize point-shaped or spherical microphones. They differ in the geometry of arrangement of the microphones and, in this respect, also in the formulae for evaluation of the coordinates as well as for corrections of the measurement error. A common feature in the design is to move the microphones away from the operating space as far as possible so that they would not stand in the way when a measurement is carried out. The possibility of introducing as large a compensation as possible of atmospheric influences is considered at the same time.

As in the case of the measurement shown in Fig. 4.16b, a relationship between the coordinates of two points is obtained; a real and a virtual one, symmetrical to one another by a plane determined by the transducer coordinates R_x, R_y, R_z. In order to obtain unambiguous results, we usually limit our calculations to a single half-space. A spatial set-up of transducers (shown in Fig. 4.18) provides us with such an arrangement. In this set-up, the transducers are arranged in a vertical plane which allows access to the operating space.

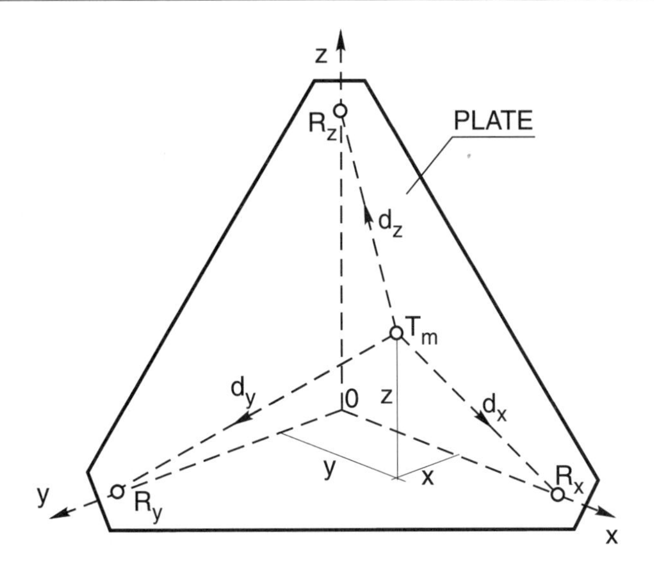

Fig. 4.18 ***A pick-up of spatial coordinates with four microphones, arranged on a plate***

4.3.3 Further possibilities of dimensional measurements

The focusing and deflection of an ultrasonic beam enables ultrasonic techniques to be useful in identifying the presence of objects, their shape, orientation and in distinguishing their movement, etc. This field may be referred to in a simplified way as ultrasonic imaging. Although simple images are often the ones in question, and they are by far not as detailed as those obtained using optical imaging, this kind of imaging has certain specific features which enable its application even under conditions where optical imaging is impossible. It is, for instance, possible to look through non-transparent materials. A wide branch of medical imaging (scanning) methods belongs here, as well as ultrasonic defectoscopy (Non Destructive Testing, NDT). This is described separately in Chapter 5. In industry, ultrasound finds its place in robotics, in imaging scenes and identifying obstacles.

An ultrasonic image differs from an optical one by the fact that it supplies information not only about the presence and shape of the object, but also about its distance. Thus the object is illuminated by the transmission of ultrasonic waves. Thus the sensing takes place in an active way when ultrasonic waves are emitted by a transducer and received subsequently after their reflection from the object. If the abilities to be focused and deflected and of penetration of an ultrasonic beam through non-transparent materials are added, we obtain a variety of innumerable applications. With the developments and advances in ultrasonic transducer technology and in measuring methods, a significant broadening in the scope of ultrasonic techniques can be expected.

As examples, two applications are described in this section which are on the borders of this scope. In Fig. 4.19, the case of sensing the presence of a screw in the observed assembly is illustrated. A short series of pulses (a burst) is emitted towards the surface to be examined, and the time of its return is evaluated. If the screw is present, the reflected pulse is detected after a time t_a, which is proportional to the distance of the screw head from the active transducer surface (in Fig. 4.19a). If the screw is absent, a pulse reflected from the bottom of the bored hole appears, in the time $t_c > t_b$ (Fig. 4.19b). A series of pulses (a so called radiopulse) is used mostly in connection with the use of piezoceramic transducers which retain some inertia on the leading and the trailing edges of pulse oscillations.

A further example represents the use of a matrix of piezoceramic transducers for measuring the shape and orientation of an object placed above the matrix (Fig. 4.20). The presence of the object is identified by the presence of reflected ultrasonic waves. In this case, the effect that, at a boundary of two different materials, each of them having different velocity of sound propagation, a partial reflection occurs, is exploited. By sequential excitation of the individual transducers and detection of the reflected waves, the position as well as the orientation of the object in a plane is evaluated. In addition the height of the object can be determined also, if the time of reflection of the wave from the upper object surface is measured and evaluated (see the detailed drawing in Fig. 4.20b).

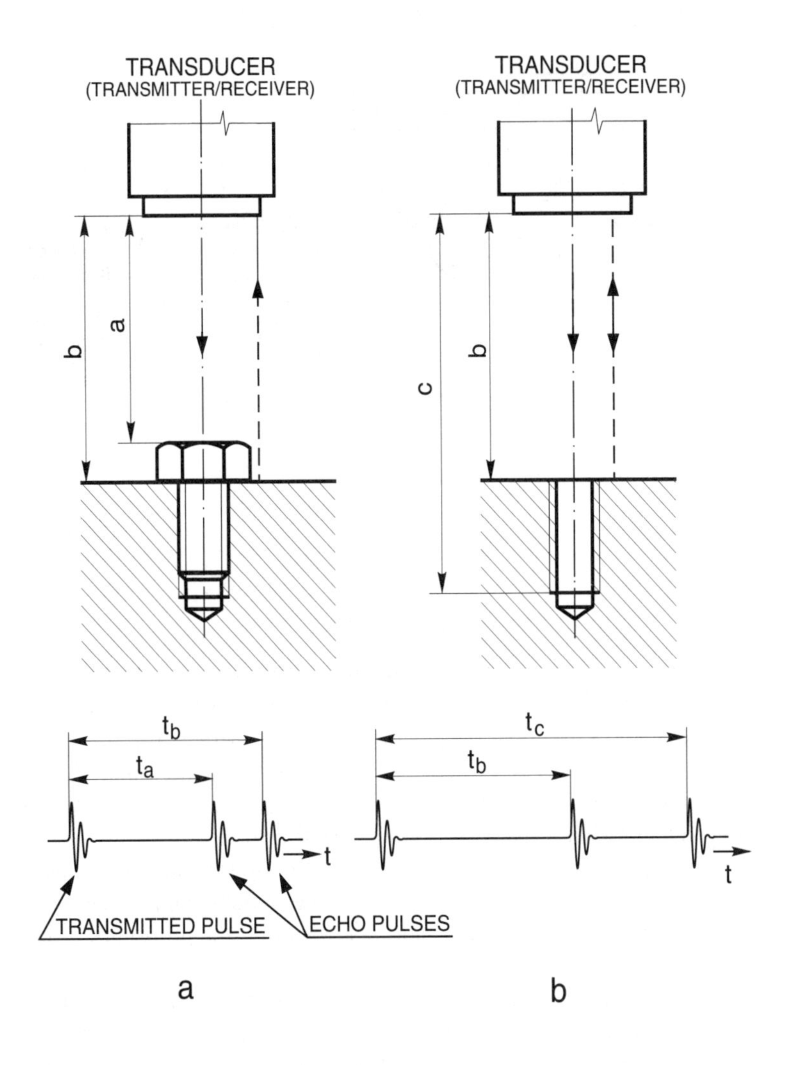

Fig. 4.19 ***Identifying the presence of an object (a screw head in this particular case) a) present, b) absent.***

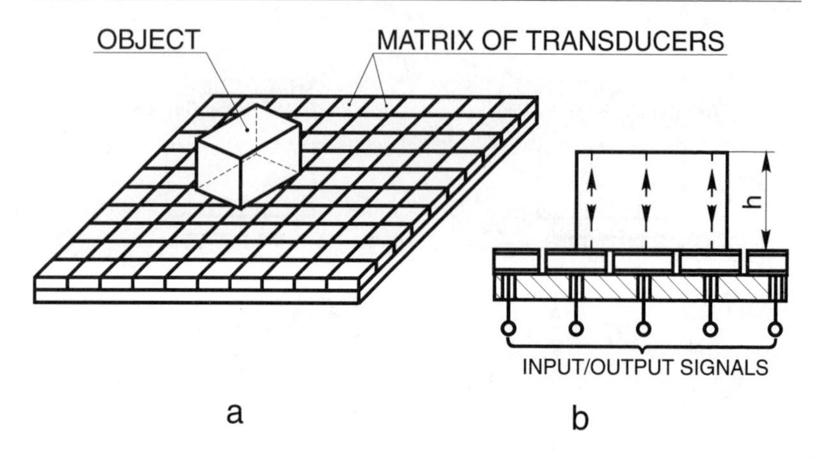

Fig. 4.20 ***Measuring the position and the height of an object using a transducer matrix***

4.3.4 Some further methods of increasing the precision of a measurement

In previous parts of this chapter attention was devoted to the ways of reducing the influence of variations of the atmospheric parameters. During a measurement, further random and systematic errors arise, some of which can also significantly affect the precision of a measurement. In many cases, computer technology, integrated into the measuring instrument, can be used to advantage to reduce these errors.

More sources of error occur in the case of pulse measurement. Therefore our attention will be devoted to this mode. When using pulse measurements, a fundamental requirement is to measure the propagation time of the ultrasonic pulse as precisely as possible, from its origin till its incidence on the active receiver surface. Thus, the task is divided into two parts: a precise determination of the start of the pulse, and a precise determination of the end of the time measuring pulse, which opens the gate for counting the clock (time mark) pulses using a counter.

Determination of the start of an ultrasonic pulse depends on the type of pulse wave generator. If a miniature spark gap is used as a point source of an ultrasonic spherical pulse wave, one of the error sources is the delay of the instant of spark occurrence with respect to the electric trigger pulse. If the time measuring pulse is started by this trigger pulse then counting may start in advance, before the actual ultrasonic pulse has appeared. This delay can arise due to the capacitance of the leading cable, which connects the voltage pulse to the spark gap electrodes. The delay is rather small as a rule, and is practically constant. Thus it can be compensated using software, or in an electronic way, if necessary. For example, the instant of spark breakdown can be detected by detecting the change of the magnetic field close to the spark gap. Since the current rise rate, di/dt, (and consequently, the rise of voltage too) is very steep, a miniature aerial is sufficient as a detector.

A similar approach can be applied when using a capacitance transducer as a transmitter. Mostly transducers of a thin, dielectric, metallized foil are used, so that their capacitance may cause a delay in the rise time of an ultrasonic wave pulse. If the delay can be determined, it is easy to eliminate it.

When using piezoceramic transducers, the remedy is a bit more difficult. Due to the large difference between the acoustic impedances of air and of the transducer, reflections and successive oscillations occur inside the transducer. This results in a slowed and delayed rise and fall of the transmitted ultrasonic waves. One possible remedy of this problem is to use a series of pulses and more sophisticated processing of the detected signal.

A precise determination of the instant of incidence of the ultrasonic pulse front face upon the active receiver surface is rather more demanding, principally for two reasons. Firstly, because the acoustic noise of the surrounding medium also acts on the receiver, and secondly, because of the intensive pulse attenuation during its propagation. The pulse amplitude varies depending on the transmitter-receiver distance. Moreover, if a spark is used for generating the ultrasonic pulse, the amplitude can even vary randomly.

The influence of the environmental noise can be rejected in such a way that the instant of termination of the measuring pulse (its trailing edge) is preset by adjusting a comparator (a Schmitt trigger) to a level which is comfortably higher than the noise level. However, this leads to a rise in the error of determination of the end of the measuring pulse, due to variations in the amplitude of the detected ultrasonic pulse (Fig. 4.21). This error can be rejected in several ways. If the dependence of this error on the duration of pulse propagation can be measured, a correction can be introduced, using software for example. In principle, there is also a simpler way which depends on a time-dependent increase in the amplification in order to obtain an approximately constant pulse amplitude at the amplifier output. Then the amplitude is relatively independent of the transmitter-receiver distance.

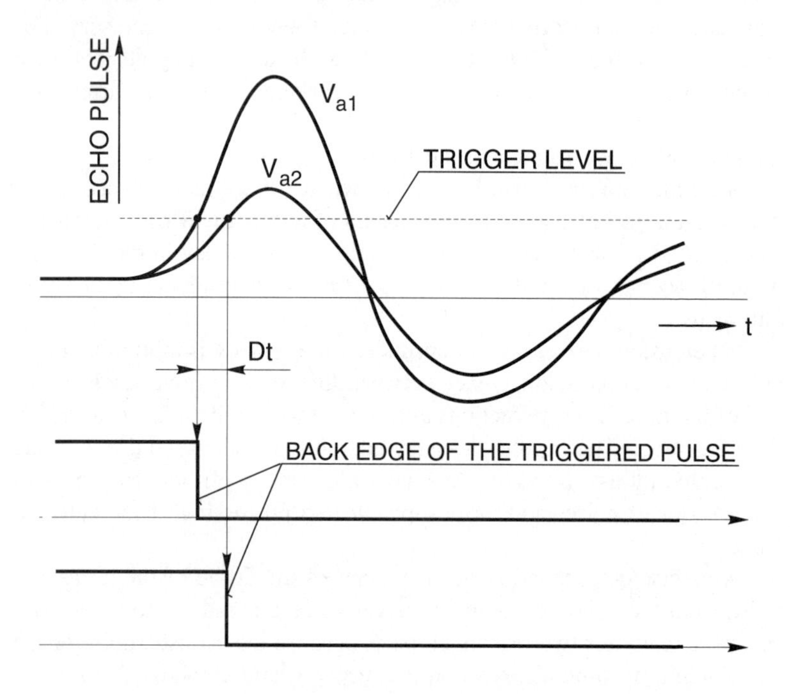

Fig. 4.21 ***An illustration of the error in determination of the instant of incidence of the ultrasonic pulse front face***

If the dependence of the amplitude of a detected ultrasonic pulse on distance is known, an amplifier with a time-dependent amplification can be used. However, it is recommendable to express the dependence using a graphical plot, or better still, as a mathematical relationship. In Fig. 4.22, one of the possible ways is indicated. A suitable non-linear voltage-dependent resistor is inserted in front of the input of an operational amplifier (a FET in this case), and a voltage from a sawtooth-shape voltage generator is connected to its gate. The generator is initialized by the triggering pulse, simultaneously with the generator of the ultrasonic pulse [ref. 4.11].

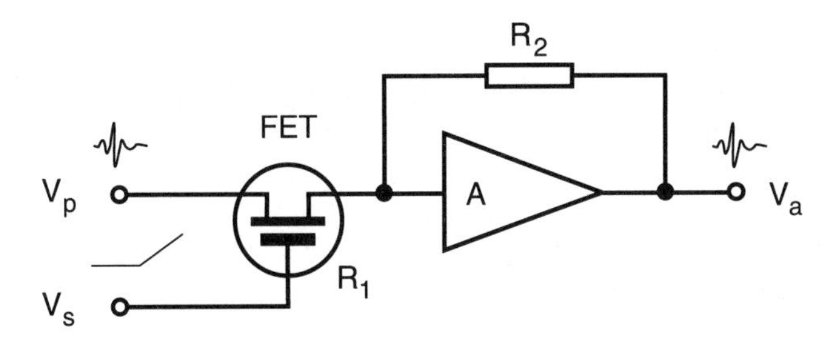

Fig. 4.22 ***A circuit for regulation of amplitude of the amplifier output pulse***

Another approach is demonstrated in Fig. 4.23. A delay line is used, in which the evaluation of the instant of termination of the time-measuring pulse is shifted by a known constant delay value, τ. Afterwards this value is subtracted. The circuit ensures that the bistable is triggered always by the same time interval of the detected ultrasonic pulse. The detector of the amplitude, V_a, of the amplifier output stores the peak voltage value and this voltage is reduced by a subsequent divider in the preset ratio (in the illustrated case by one half of the original amplitude). This voltage value is then subtracted from the voltage at the output of the delay line. The voltage difference, V_d, is led to the comparator input which changes its output state at the instant when the voltage, V_d, crosses zero. This complete circuit is connected between the amplifier output and the bistable input, according to Fig. 4.23.

Rejection of the environmental interference can be achieved by inserting a filter into the amplifier circuit. The narrower the transmission band of the filter, the more efficient the noise rejection. However, a high quality transducer is necessary in this case as well as a sufficiently stable value of the frequency components of the received signal.

4.4 Ultrasonic measurement of liquid flow velocity

In recent years, many types of various ultrasonic flow gauges have been designed and invented. Their attractiveness rests in the fact that they do not hinder the liquid flow, and therefore no losses arise in measurement. They are not demanding of regular maintenance and the liquid viscosity plays no role. As the operation of these flow gauges is independent of the

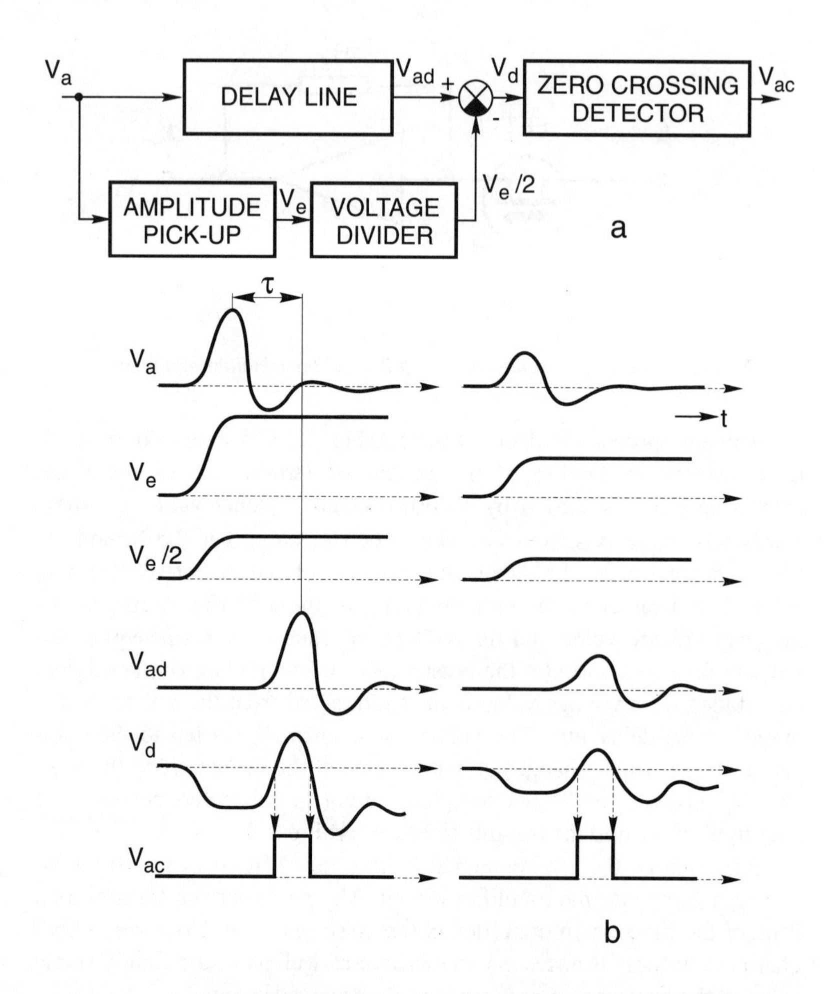

Fig. 4.23 ***A circuit for regulation of the level of a comparator output signal clamp***

velocity of sound in liquid, they can be used practically over a wide range of values of working temperature.

Ultrasonic measuring methods for liquid flow velocity are based on the mutual interaction of a medium with the medium flow. This method has advantages such as contactless measurement, independence of the conductivity of the medium, no pressure losses in the pipeline, and so on.

One drawback is the considerably more complicated electronic circuitry required for processing the information-carrying signal.

As shown in Fig. 4.24, 5 methods can be used for measurement in principle, which differ from each other by the method of processing the information-carrying quantity.

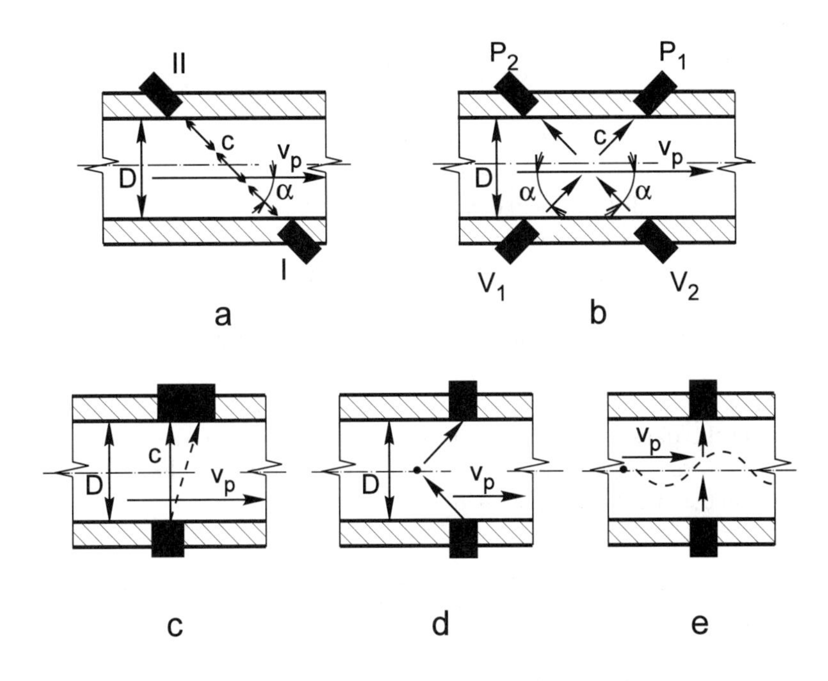

Fig. 4.24 ***Illustration of ultrasonic methods for measuring liquid flow***

In the first method, the measured signal is obtained from the change of the propagation velocity of ultrasonic waves, caused by the flow of the medium. The resultant velocity of propagation of the ultrasonic waves, in a place where the medium flows with a velocity v_p, is given by a vectorial sum of the propagation velocity, c, in the non-flowing medium, and the velocity v_p. The ultrasonic wave propagates alternately from transducer I to transducer II, and vice versa. The resultant velocity of the wave is

in the first case: $c - v_p.\sin\alpha$

in the second case: $c + v_p.\sin\alpha$.

These changes in the propagation velocity are evaluated by measuring its transition duration through the medium. In order to reduce the effects of temperature and of medium composition to a minimum value, the time difference is evaluated both in the direction of flow and in the opposite (antiparallel) direction

$$\Delta t = \frac{2l}{c^2}.v_p.\sin\alpha \tag{4.30}$$

where c - velocity of ultrasound propagation
v_p - medium flow velocity
l - distance between transducers
α - angle between the directions of the ultrasound wave propagation and the medium flow.

The variant b represents a similar case, the only difference rests in different processing of the signal.

For the alternatives a and b, where a phase angle is calculated following the transition of the ultrasonic wave motion through a medium, the following expression holds true for propagation of the ultrasonic wave in the direction parallel with flow:

$$\varphi_1 = \omega t_1 = \omega \frac{l}{c + v_p \sin\alpha} \tag{4.31}$$

and in the antiparallel flow direction

$$\varphi_2 = \omega t_2 = \omega \frac{l}{c - v_p \sin\alpha} \tag{4.32}$$

where l - distance between probes
t_1,t_2 - time of travel of the ultrasonic wave motion in the parallel and antiparallel direction
c - velocity of wave propagation in the stationary liquid
v_p - average flow velocity of the medium
$\omega = 2\pi f$ - angular frequency of ultrasound.

The phase difference between both signals is given by:

$$\Delta\varphi = \varphi_2 - \varphi_1 = \frac{2\omega l.\sin\alpha}{c^2 - v_p^2.\sin^2\alpha} \qquad (4.33).$$

In the case of evaluation of the frequency difference between waves propagating in both directions, with respect to the liquid flow (alternatives 1 and 2), we arrive at the expression

$$f = f_1 - f_2 = \frac{1}{t_1} - \frac{1}{t_2} = \frac{2v_p}{l}\sin\alpha \qquad (4.34).$$

The last expression, in contrast to the previous ones, is significant for the reason that the velocity of ultrasound propagation, c, is cancelled here.

For the sake of completeness, the three remaining methods should be mentioned. In the third method, c, the deflection of the ultrasonic waves caused by the flowing medium is exploited. The wave motion propagates from the transducer to the receiver in a perpendicular (i.e. shortest) direction, provided the medium is stationary. If the medium begins to flow, the beam gets deflected, and a change in the signal appears on the receiver.

This method yields only qualitative (approximate) results and this is why it has not become widespread.

The fourth method, d, is suitable for flow measurements in a medium containing solid particles from which ultrasonic waves can be reflected. It is denoted as a Doppler effect based method (a Doppler method). Due to reflection from the moving particles, the received wave has a change in its frequency compared to the emitted frequency. The method is not suitable for a medium which does not contain solid particles.

In the last method, e, a frequency superposition of so-called von Karman's curls upon the ultrasonic waves transmitted across the medium is exploited.

This simplified survey gives an idea of the measuring principles and techniques of processing ultrasonic signals. The elimination of the effect of the propagation velocity, in methods 1 and 2, favours these methods over the other ones, in which this influence is not fully eliminated. The problem is not straightforward though, because of the dead zones which need to be overcome by the ultrasonic pulse. The dead zones arise when built-in probes are used in the pipeline. The propagation time in one direction is

$$t_1 = \frac{A}{c} + \frac{B}{c} + \frac{1}{c + v_p \sin\alpha} \quad (4.35)$$

and, in the opposite direction,

$$t_2 = \frac{A}{c} + \frac{B}{c} + \frac{1}{c - v_p \sin\alpha} \quad (4.36).$$

On evaluating the frequency difference of the pulse circulation in the two directions, with respect to the flow direction, or (alternatively) the time interval, Δt, we obtain a more complex expression:

$$\Delta t = t_2 - t_1 = \frac{1}{c - v_p \sin\alpha} - \frac{1}{c + v_p \sin\alpha} = \frac{2 l v_p \sin\alpha}{c^2 - v_p^2 \sin^2\alpha} \quad (4.37).$$

It is evident that the expressions are acquiring a complicated form, though the influence of the flow velocity is not eliminated. In this case, compensation must be introduced for fluctuations of the propagation velocity caused by dependence on temperature, and on the composition of the medium. For this compensation, propagation of a pulse perpendicularly across the medium flow is utilized. In such a set-up, the influence of the flow is eliminated, and the duration of the pulse propagation depends merely on the distance between the transducers, on the temperature and on the composition of the medium. The signal obtained in this way is used for compensation of the flow-measuring signal.

Only two types of ultrasonic flow gauges satisfy the fundamental criterion that the efficiency of the instrument should not be affected by variations of the natural sound velocity in the flowing medium. One of them is a purely analogue type based on a principle of a frequency difference between the time-separated acoustic paths. Another one is an essentially digital method of calculation of a diameter (area x velocity) of several ultrasonic trajectories.

From theoretical considerations, it may be assumed that the method of the frequency difference fits better for a fully developed flow of the medium, with a Reynolds number higher than $3x10^4$, and with an average velocity lower than the natural velocity of sound. With these conditions, and within a given dynamic range, the measuring instruments employing this method are very likely to be more efficient than the common instruments that utilize and measure the pressure difference, because they are less dependent on the characteristics of an ideal liquid.

The second type of ultrasonic flow gauge is suitable for a far wider range of flow conditions, if more sophisticated electronics is used in order to obtain wider versatility and higher precision, which naturally leads to higher financial demands on the instrument.

If advantages such as electronic subtraction are considered, a minimum requirement for maintenance and the absence of pressure losses, then in this respect, ultrasonic flow meters are beginning to be successful in competition with common flow gauges, as well as on the commercial level. However, it seems that ultrasonic flow meters, in the immediate future, will become complementary to traditional flow gauge instruments rather than replacing them.

4.5 Measuring the thickness of materials

Industrial ultrasonic measurements of thickness are usually carried out as measurements of the time interval of transmission of a longitudinal ultrasonic wave pulse in the forward and, after reflection, in the backwards directions. A wave is transmitted through a sheet, a plank or a wall, and then a conversion of the measured time to a thickness value follows, for materials for which the ultrasound propagation velocity is known.

An analog or digital value is often complemented by an oscilloscope display reading, for instance in the case of determination of corrosion. The measured values are typically in the range from 0.2 mm up to 1 m in thickness, however, they are not usually obtained in a single measurement at a time. The accuracy is better than 1 %, the resolution is about 2.5 μm and the time response, for precise instruments, is about 1 s.

At present, mostly digital thickness gauges are used, implemented with integrated circuits, and battery operated. These portable instruments work with direct probes, most often doubled, which have some advantages, such as a small dead zone, more convenient application to curved surfaces and a simpler indication of point corrosion. In order to be able to measure small thicknesses, the probes are equipped with a separating layer whose thickness is a few times higher than a few multiple echoes.

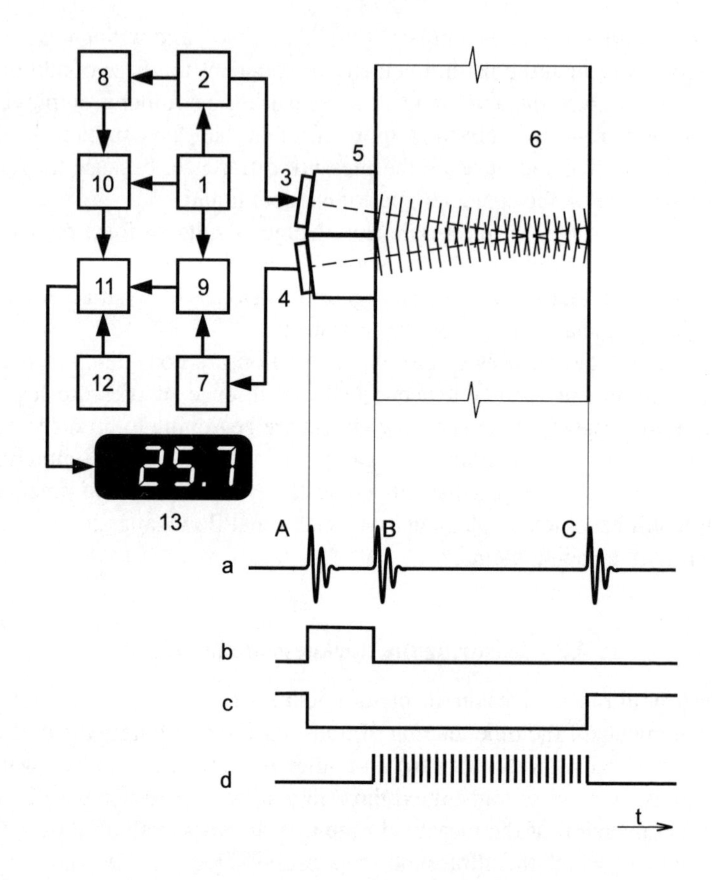

Fig. 4.25 ***A schematic drawing of an ultrasonic pulse digital thickness gauge 1 – synchronizing circuit, 2 – pulse generator, 3 – emitting transducer, 4 – receiving transducer, 5 – separating layer, 6 – measured object, 7 – amplifier, 8 – auxiliary amplifier, 9 – bistable circuit, 10 – bistable circuit, 11 – gate, 12 – oscillator, 13 – counter***

In order to eliminate the temperature dependence of the wave propagation velocity (in the separating layer), special electronic circuits and systems are used. These eliminate the dependence in such a way that a first echo is obtained from the separating layer and the second one from the measured object whereby the signal from the first echo serves as an initialization.

The principle of operation of an ultrasonic, pulse utilizing, digital thickness gauge is indicated in Fig. 4.25. The emitting transducer, 3, is excited by an electronic pulse from the generator, 2, which is controlled by a signal from a synchronization circuit, 1. The emitted ultrasonic signal, A, propagates through the separating layer, 5, to the measured object, 6. The echo, C, reflected from the rear wall of the object falls on the separating layer, 5, and it propagates through to the receiving transducer, 4. The electronic signal from the receiver is led to an amplifier, 7. The echo B from the interface of the separating layer and the measured object falls on the emitting transducer, 3, from which the signal is led to an auxiliary amplifier, 8. The amplifier outputs, 7 and 8, are connected with bistable circuits, 9 and 10. These clamp at an instant A, simultaneously with the emission of a pulse from the transducer, 3. The bistable circuit, 9, returns to its initial state at the time of arrival of the pulse B originating from the transition of the separating layer and the measured material, and the circuit, 9, after arrival of the pulse C, reflected from the bottom of the object. The pulses B and C, from the bistables 9 and 10, are led to the inputs of a gate, 11, together with the time mark (clock) pulses from a local reference oscillator. The gate realizes an inverse logical operation (the function NAND). At its output, a pulse count appears that is proportional to the material thickness, which is indicated directly on the counter display, 13.

Large automatic inspection and testing machines exist also, where ultrasound is utilized for the determination of dimensions. For example, for inspection and monitoring of sheath pipes of fast reactor fuel rods, a checking automatic machine was built which allows the measurement of dimensions with an accuracy of 1 μm, within a measuring range of 0.1-5 mm. At a translational speed of 60 m/min of the sheath pipe, the rate of data aquisition is 6 points per mm of translational shift, and in addition the pipe diameter and the pipe wall thickness are checked at the same time.

In the manufacture of plastic pipes on extruders, an ultrasonic tester proved suitable for continuous inspection of the wall thickness, and for identification of defects. The output signal obtained by the sensor is simultaneously used for regulation of the extruder in order to maintain a constant, prescribed pipe wall thickness. On both ends of the sensor chamber, face seals are fitted and the chamber is filled with bubbleless water. The bubbleless water represents a coupling medium between the sensor and the inspected pipe. The transfer of a signal from the rotational part to the stationary block is secured by cylindrical windings of a special transformer. The rotor position is traced by a proximity sensor via a toothed wheel, and the signal obtained in this way is further processed in the electronic control system.

4.6 Temperature and pressure measurement

4.6.1 Temperature measurement

Ultrasonic thermometers are usually designed to be sensitive to the temperature dependence of the sound velocity, c. Two elementary principles are used: resonance and pulse-based.

The dependence of the natural resonance of bending oscillations of a small disc quartz resonator was utilized in a product of the Hewlett-Packard Company. The probe of instrument HP 2801A operates at a frequency of 28 MHz and its temperature sensitivity is 1000 Hz/°C, in the temperature range from -40 °C up to 120 °C.

For measurement of high temperature, a pulse sensor is used. Its fundamental part is a resonator operating as a temperature pick-up. Ultrasound is emitted into the resonator, from a magnetostrictive transducer, via a transfer line of thin wire. On an oscilloscope, the state of resonance of the pick-up can be observed during continuous tuning of the exciting oscillator. The measured temperature is determined from a calibration curve, which indicates the dependence of the resonance frequency on temperature. When using a ruthenium resonator, a sensitivity of 20 Hz/°C was obtained at the working frequency of about 136 kHz. These thermometers enable temperature measurements up to 1800 °C [4.10].

Another alternative application of this method is the measurement of high temperatures above 2000 °C inside nuclear fuel rods [4.11]. The temperature dependence of the pulse propagation time through the fuel rod is utilized in this case.

In gases, the temperature can be measured directly, because the velocity of propagation, c, is determined by the gas composition and the temperature. It can be calculated from the equation

$$c = (\gamma R\Theta/M)^{1/2} \tag{4.38}$$

where γ - ratio of the specific heats c_p/c_v

R - molar gas constant (8.3 JK^{-1})

Θ - absolute temperature
M - mean molecular weight.

For the measurement described above, a pulse principle is used. An advantage of this method is that it has no inertia as the measured quantity is proportional to the mean of the square of the velocity of the gas molecules. On the other hand, a drawback of this method is that the precision of the measurement is influenced to a considerable extent by the turbulence of the gas flow.

When carrying out a measurement, the gas flow velocity must be eliminated. Therefore, propagation of a pulse in the parallel, and simultaneously in the antiparallel direction, with respect to the gas flow, is used for measurement. In studying the transition properties of gases, temperatures in the range 10 000 °C to 20 000 °C were measured by the ultrasonic pulse method. Then the gas itself served as a pick-up at the same time.

4.6.2 Pressure measurement

The propagation of sound in gases, liquids and solids is influenced by pressure in various ways. In solids, the pressure causes a so called tension-induced anisotropy. In liquids, the influence of pressure on the propagation velocity, as well as on the absorption, is small. For example, the depth pressure effect (increase of sound velocity in sea water with increasing depth h, due to hydrostatic pressure) is $\Delta c/\Delta h = 0.017\ s^{-1}$.

In gases, the velocity of sound propagation, c, increases with increasing pressure, p, while the absorption, α, has an inverse dependence on pressure, p.

Among the commercially produced instruments for pressure measurement there belongs, for example, a pressure gauge from the Hewlett-Packard Company, type 2811 A, where a special section of quartz is used as a pick-up.

The response of a quartz crystal to a pressure, p, is almost linear; however, it does need temperature compensation. The probe of the instrument contains a quartz oscillator, which senses the pressure, and a reference oscillator. The oscillator frequency changes from 0.5 to 1 MHz. The pressure range is from 0 to 84 MPa, with a resolution of 70 Pa.

Another principle is based on measurement of the height of a mercury column. A piezoelectric transducer, placed at the bottom of a mercury column, emits ultrasonic pulses to the mercury surface and senses the reflected echo from the surface. If there are two transducers in two mercury columns, whereby one of them is connected to the measured space and the other one to a standard ambient atmosphere, the time of propagation depends on the height difference of the two columns, provided the propagation velocity is constant. The frequency of the clock oscillator can be adjusted in such a way that the result can be read in any arbitrary units.

Maintaining a constant velocity of sound requires consequent regulation of temperature. For an accuracy of measurement of ±1.625 Pa, regulation of the temperature to better than ±0.02 °C must be ensured. The measurement range of this type of pressure gauge is from 0 to 104 000 Pa (i.e. up to ~1 atm). It is suitable for calibration of pressure transducers and gauges, continuous measurement of pressure in vacuum chambers and so on. One manufacturer of such instruments, which are commonly accessible on market, is the company Wallace Tiernan Div.

4.7 References

4.1 *Moritz W.E., Shreve P.L. Mace L.E.:* Analysis of an Ultrasonic Spatial Locating System, IEEE Transactions on Instrumentation and Measurement, 25, 1976, No. 1, 43-50

4.2 *Merhaut J.:* Teoretické základy elektroakustiky, Academia, Prague, 1985, 385p. (in Czech)[1]

4.3 *Morfey C.L., Howell G.P.:* Speed of Sound in Air as a Function of Frequency and Humidity, J. Acoust. Soc. Amer. 68, 1980, 1525-1527

4.4 *Obraz J.:* Ultrazvuk v měřící technice, SNTL, Prague, 1984, 484p. (in Czech)[2]

4.5 *Kay L.:* Airborne Ultrasonic Imaging of a Robot Workspace, Sensor Review 5, 1985, No. 1, 8-12

4.6 *Ahrens U.:* Solutions and Problems in Applications of Airborne Ultrasonic Sensors in Assembly- and Handling System, Robotersysteme I, 1985, No. 1, 17-20

[1]Theoretical principles of electroacoustics

[2]Ultrasound in measuring techniques

4.7 *Commercial press:* Prospectusses of Sonic Digitizers, SAC, (Science Accessories Corporation, Strattford, USA)
4.8 *Zenin V.J., Maslyukow V.A., Sych V.P.:* Ustroystvo dlya schityvania graficheskoi informacii, Pat. USSR No. 525976 (in Russian)[3]
4.9 *Kočiš Š.:* Improvement of Accuracy of Acoustical Digital Coordinate Measurement, Preprint, In: ACTA IMEKO 1979 (VIIIth World Congress IMEKO '79), 409-414
4.10 *Moritz W.E., Shreve P.L.:* A Microprocessor-Based Spatial-Locating System for Use with Diagnostic Ultrasound, Proc. IEEE, 64 (1976), No. 6, 966-974
4.11 *Kočiš Š.:* Ultrasonic Pulse Method and its Application for Numerical Plane and Space Measurement, Elektrotechnický časopis, 35, 1984, No. 11, 480
4.12 *Pomeroy S.C., Dixon H.J., Wybrow M.D., Knight J.A.G.:* Ultrasonic Distance Measuring and Imaging Systems for Industrial Robots, Robot Sensors 2, Springer, 1986, (see also: Robotics 3, 1987, 181-188)
4.13 *Ulrich M., Lazaro S.:* Why Ultrasonics for Noncontact Position Sensing, Instrumentation and Control Systems 5, 1992, No. 4, 39-42
4.14 *Figeroa F., Barbieri E.:* An Ultrasonic Ranging System for Structural Vibration Measurement, IEEE Transactions on Instrumentation and Measurement 40, 1991, No. 4, 764-769
4.15 *Fiorillo A.S.:* Design and Characterisation of a PVDF Ultrasonic Range Sensor, IEEE Transaction on Ultrasonics, Ferroelectrics, and Frequency Control 39, 1992, No. 6, 362-367
4.16 *Hickling M., Marrin S.P.:* The Use of Ultrasonics in Air, J. Acoust. Soc. Amer. 79, 1986, No. 4, 124-127
4.17 *Shimokohbe A., Ma S.:* Ultrasonic Measurement of Three-Dimensional Coordinate, Preprint, In: Proc. XIth IMEKO World Congress, 441-447
4.18 *Sano S.:* Mobility Aids for Blind, In: Electronic Devices for Rehabilitation, Chapman & Hall, London, 1985, 79
4.19 *Teshigawara M., Shibata F., Teramoto:* High Resolution (0.2 mm) and Fast Response (2 ms) Range Finder for Industrial Use in Air, In: Proc. IEEE Ultrason. Symp., 1989, 639-642
4.20 *Joffe D.:* Polaroid Ultrasonic Ranging Sensors in Robotic Application, Robotic Age, 1985, 23-30
4.21 *Moosby E.G.:* Practical Ultrasonic Thermometer, Ultrasonics 7, 1969, No. 1, 13-15
4.22 *Tasman H.A., Patzold E.E.:* Ultrasonic Thermometer, 1978, U.S. Patent, No. 4.195.523
4.23 *Commercial press:* Quartz Pressure Gauge, Typ 2811A, Hewlett-Packard, Ltd

[3]A device for pick-up of graphical information

5 Non-destructive testing (NDT)

Testing of materials is important, primarily in a field of increasing quality and reliability of products. NDT finds application especially in the machine industry. Non-destructive methods occupy a special position because they ensure that the material and its mechanical properties are not damaged. The existing non-destructive methods are based on the principle of detection of changes caused by inhomogeneities inside the material (fissures, ruptures, shrinkage cavities, holes, inclusions, etc), either in magnetic fields, eddy currents, optical changes or, last but not least, acoustic waves.

Ultrasonic methods have their advantages due to their capability to precisely determine position and, if need be, also the shape of a defect. Ultrasonic defectoscopy (NDT) enables recognition of defects arising in the manufacturing process, or during an operational period, as a result of mechanical fatigue.

In this chapter, the fundamental principles of ultrasonic defectoscopy are introduced, as well as methods of testing and their most widespread applications.

5.1 Basic principles of ultrasonic NDT

In Table 5.1 the basic principles of ultrasonic non-destructive testing are presented. Method A represents a so called transmission method which is based on the effect that an ultrasonic shadow is created behind an obstacle which is represented as an inhomogenity inside the material.

The pulse reflection method, indicated in diagram B, is the most widespread among the methods of non-destructive testing of materials by ultrasound. It provides plenty of information about the examined object. It exploits the scanning effect of an ultrasonic pulse, reflected from an inhomogeneity. In diagram C, a resonance method is indicated which uses standing waves in a material. If there is a change of resonance, referenced to an undamaged material, it is possible to infer the presence of an inhomogeneity.

TABLE 5.1

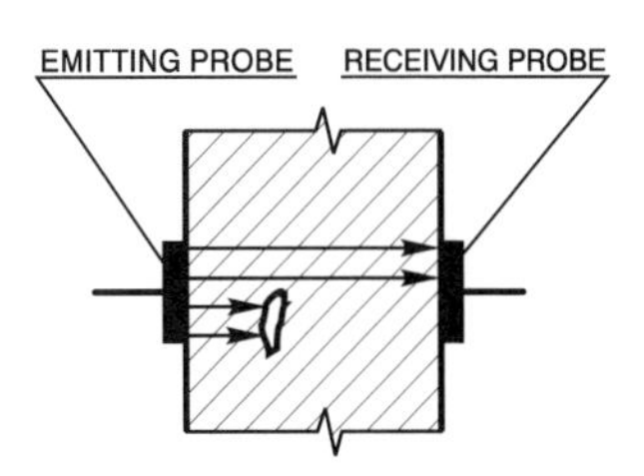

a TRANSMISSION METHOD

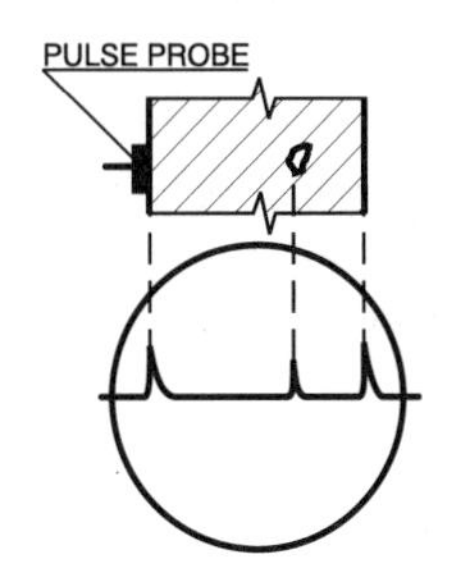

b PULSE REFLECTION METHOD

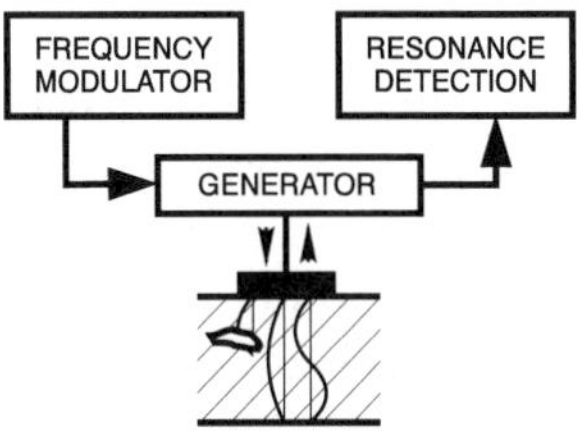

c RESONANCE METHOD

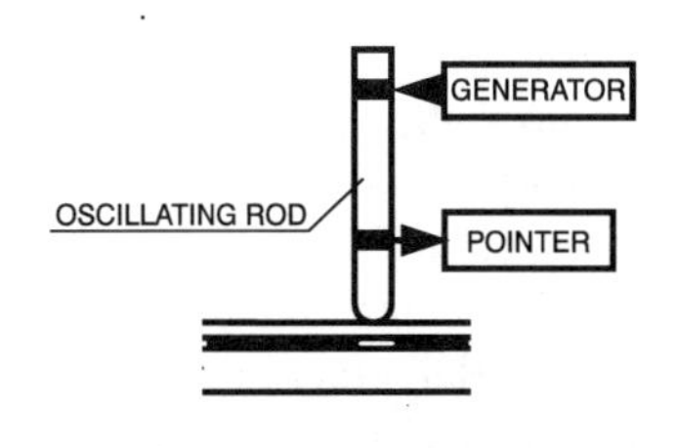

d IMPEDANCE METHOD

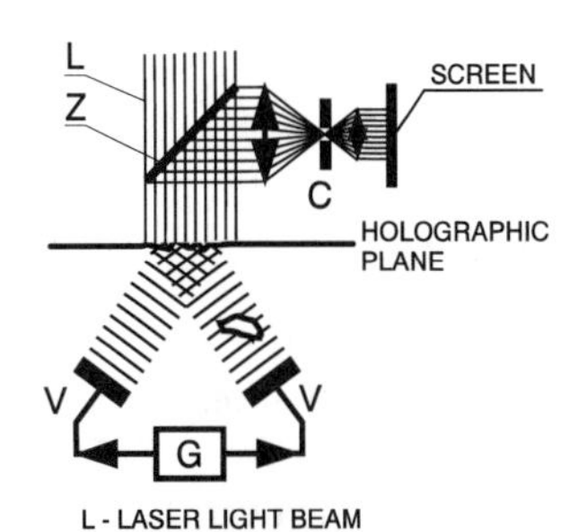

e HOLOGRAPHIC METHOD

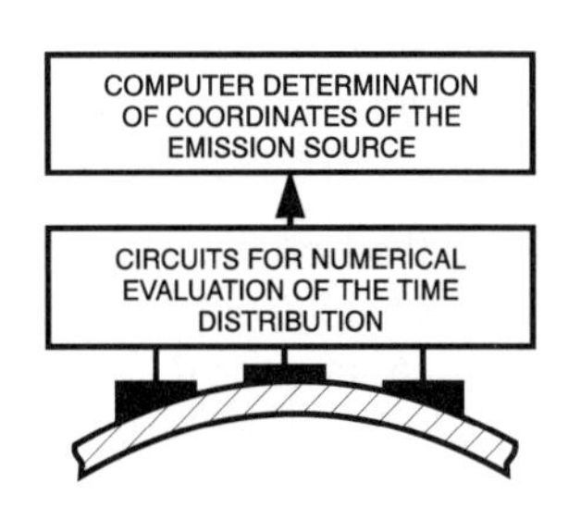

f ACOUSTIC EMISSION METHOD

The impedance method, D, utilizes a difference in the acoustic impedances between an undamaged and a damaged piece of the examined material.

The holographic ultrasonic method is indicated in diagram E. It uses imaging based on interference of the ultrasonic waves. The method of acoustic emission, F, is based on an analysis of the signals which arise at the broadening of a rupture in the examined material.

5.2 Ultrasonic methods in NDT

5.2.1 Transmission method

The transmission method is based on the propagation of an ultrasonic wave from a transmitter to a receiver, both of which are placed on the opposite surfaces of the material to be tested. If there is no inhomogeneity between the transmitting and the receiving probes, a given pressure acts upon the receiving probe. In the case of the presence of an inhomogenenity between the probes, a drop in the acoustic pressure appears. A drawback of this method is that the drop in the acoustic pressure can also occur by a faulty coupling of the probes to the material, or by a change in the material structure. The method can be applied with advantage in an immersion mode of the probe-to-material coupling.

The transmission method is characterized by a total absence of dead zones. The waves can be emitted continuously, or in a pulse mode. In spite of simple processing, the transmission method is used only in special cases of non-destructive testing.

5.2.2 Pulse reflection method

This method provides the largest amount of information about the examined material, either from the point of view of the inhomogeneities, or of their structure and dimensions. The principle of operation of the reflection method is given by Fig. 5.1. The probe, 3, is excited by a generator of transmission pulses, 2, which is controlled and synchronized by the timing generator (timer), 1. The probe, 3, is acoustically coupled to the inspected medium, 4. The ultrasonic pulses, reflected from the inhomogeneities at the rear side of the inspected material, are returned to the probe, 3. At the same time, a generator of a sawtooth shape voltage, 7, is triggered by the timer, 1. The output sawtooth-shaped pulses are used as a time base for the horizontal deflection plates of the CRT, 6. The amplified echo pulses from the output of the amplifier, 5, are passed to the vertical deflection plates.

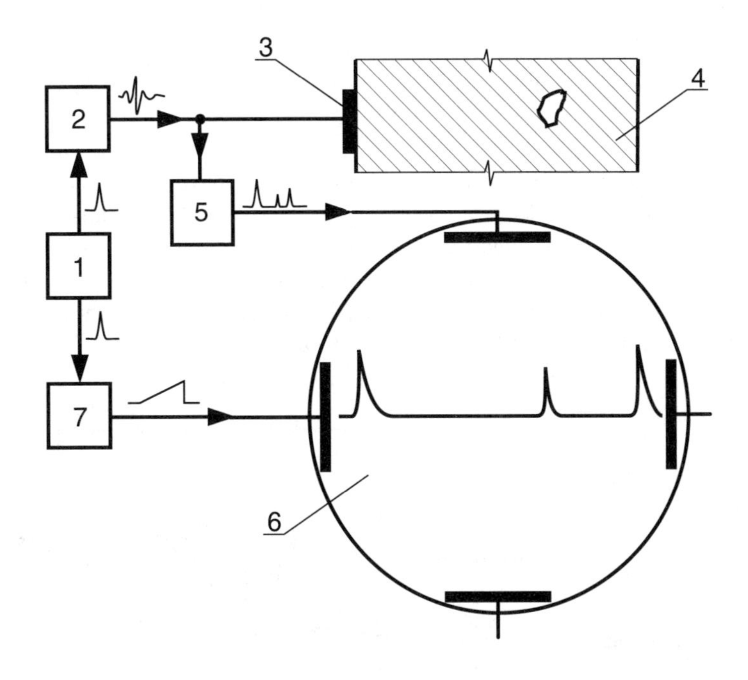

Fig. 5.1 ***Schematic circuit diagram of a pulse defectocope***
1 - timing generator (timer), 2 - emitted pulse generator, 3 - ultrasonic probe, 4 - test medium, 5 - amplifier, 6 - CRT, 7 - sawtooth voltage generator

The pulse reflection method provides information about the distance of the reflecting surface, according to the delay with which the reflected pulse returns to the probe. The size of the reflecting surface is characterized by the echo height. The pulse shape is distorted according to the properties of the reflector, or the medium. By analysing the pulses in the frequency domain, we obtain information about the character and the orientation of a defect.

5.2.3 Resonance method

The application of the resonance method is limited. It has found its practical use limited to testing objects of a constant thickness (e.g. rails). The ultrasound frequency is preset so that the measured thickness equals an integer multiple of a half-wavelength, given by the relationship

$$d = n\frac{\lambda}{2} = n\frac{c}{2f_r} \quad (n = 1,3,5...) \qquad (5.1).$$

This state of resonance shows up in the electroacoustical transducer inside the probe as a decrease of its input impedance. If the transducer, 1, is connected into an electrically tuned oscillator circuit (Fig. 5.2), the current, i, increases and flows to the transistor, 3, through the milliammeter, 4. The resonance frequency, f_r, is found, using the tuning capacitor, C or, if need be, the variable inductance, L.

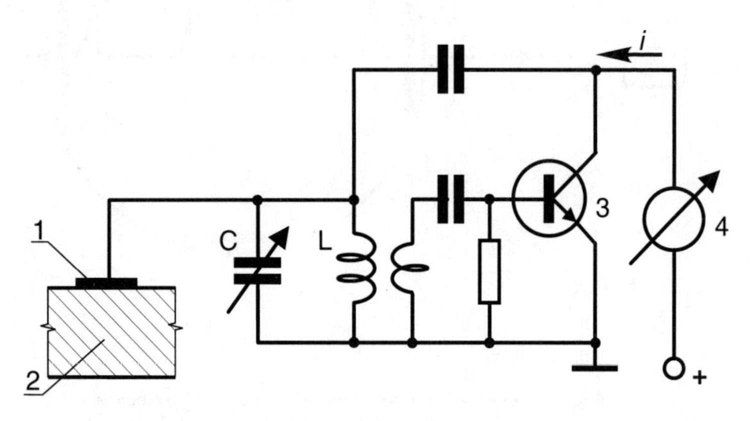

Fig. 5.2 ***Circuit diagram of a resonance instrument***
1 - ultrasonic probe, 2 - test medium, 3 - active element of the oscillator, 4 - ma meter

For this measurement, it is necessary to eliminate the unknown harmonic frequency number. This can be accomplished by two measurements at the nearest neighbour frequencies, f_n and f_{nx}. The base harmonic frequency, f_r (for n = 1) is obtained as the difference of frequencies,

$$f_r = f_{n+1} - f_n .$$

By substituting into (5.1), the required thickness can be calculated.

If the resonance condition is disturbed, it is possible to conclude that there is an inhomogeneity below the probe.

When using the electrical tuning of frequency, the method can be automated, according to Table 5.1.c. The signal from the generator is frequency-modulated by a sawtooth voltage, and is proportional to the instantaneous value of frequency, and the sweep signal for the time base varies as well. The time base signal is led to the horizontal deflection plates of the CRT monitor screen. From the screen the position of the overshot ringing, (clicks), or pulses is read, which arise at the resonance frequency of the measured medium, limited by the outer wall of the material, or by an inhomogeneity. The mutual distance of individual clicks can also be converted to a voltage, and this can be measured by a voltmeter.

5.2.4 The impedance method

In the impedance method, the difference in acoustic impedances between an undamaged and a damaged piece of the inspected medium is exploited. The measured quantity is the amplitude or phase of a reactive force, F, which is related to the impedance, Z, through a relation

$$F = c.Z \qquad (5.3),$$

where c is the acoustic velocity.

The acoustic impedance is a complex value and it is possible to measure its modulus, $|Z| = \sqrt{Re^2\{Z\} + Im^2\{Z\}}$,

and its phase, $\varphi = Im\{Z\}/Re\{Z\}$.

The impedance method is used mostly for checking the quality of connection of layers. Mechanical oscillations are introduced into the medium to be tested and their frequency ranges from 1 kHz up to 50 kHz. The space below the probe is forced to oscillate mechanically as a whole. The changes in the mechanical impedance are exhibited in a backward response of the force, F, acting on the piezoceramic transducer inside the probe. Either a ratio of the voltage amplitudes, U_2/U_1, on the emitting and the receiving transducers is compared, or the phases between these two voltages. In the case of disconnection, or faulty connection of the layers, the impedance changes because the material below the faulty connection does not oscillate.

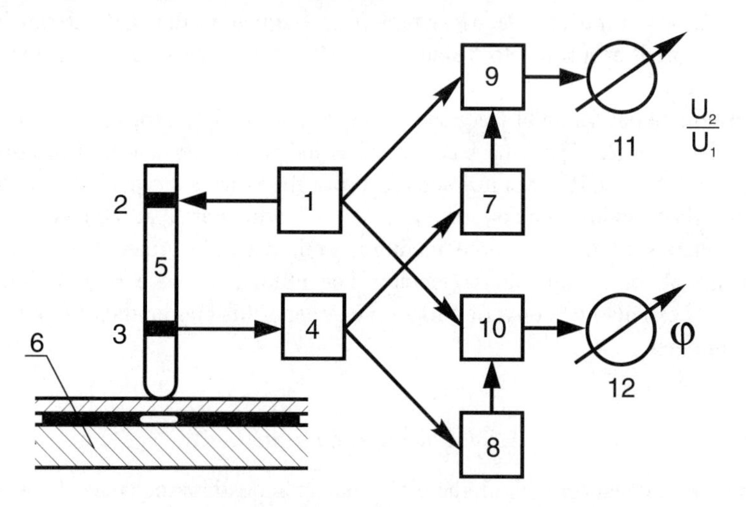

Fig. 5.3 ***Schematic diagram of an impedance defectoscope***
1 - generator, 2 - emitting transducer, 3 - receiving transducer, 4 - preamplifier, 5 - resonant rod, 6 - test material (medium), 7,8 - amplifiers, 9 - amplitude detector, 10 - phase detector, 11,12 - pointer deflection indicators

In Fig. 5.3, a schematic diagram for the identification of disconnections in multilayer materials is indicated. The emitting transducer, 2, is excited by a generator, 1. The transducer emits continuous oscillations by which a bar, 5, is forced to oscillate. The oscillations of the bar are detected by a receiving transducer, 3, and their amplitude as well as their phase depends on the medium below the bar. If there is a disconnected layer below the bar, a change in the force arises which acts on the receiving transducer, 3. Its output signal is led to a preamplifier, 4, and from its output, the signal branches into two parallel channels, 7 and 8, whereby in circuit element 9, the amplitude is compared with the transmitted signal and, in circuit element 10, the phase is compared with the phase of the emitted signal. The changes are indicated by deflection meters, 11 and 12.

5.2.5 Holographic method

Holography is based on interference phenomena [5.1], which arise in a holographic plane by superposition of the front faces of two waves: a wave $A_{(j\omega)}e^{j(\omega t+\varphi)}$, containing information about the imaged object and a reference wave $B_{(j\omega)}e^{j(\omega t+\varphi)}$. The expressions $A_{(j\omega)}$ and $B_{(j\omega)}$ are complex functions of coordinates in a holographic plane. The record of an image in a holographic plane is called a hologram. A necessary condition for its creation is coherence of both waves. A hologram contains full, three-dimensional information about the imaged object but it does not provide a direct picture. It can be recorded on a photographic film. If a developed film - a hologram - is illuminated using a laser whose light was originally used as a reference wave, a three-dimensional picture is obtained. An advantage of holography is the capability of obtaining a three-dimensional picture while the hologram itself is 2-dimensional.

Acoustic holography, which allows one to display the internal space of a medium, is based on the principle described above. Two ultrasonic radiators are used whereby the waves from one of them pass through the material under investigation, while the other one serves as a reference [5.2]. The condition of coherence can, in the case of ultrasonic transducers, be satisfied relatively easily because the emission is possible only within a narrow frequency band.

In Table 5.1, the diagram e illustrates an arrangement for the creation of a hologram on a liquid surface. Upon incidence of both waves on the liquid level, i.e. the reference wave and the wave having passed through the examined object, an interference pattern arises on the liquid surface. It gives rise to an interference picture - a hologram. Both of the emitting transducers, V, are supplied and excited from the common generator, G, which generates a continuous wave oscillation with a certain modulation (sweeping of the frequency, or chirping), so that the creation of standing waves is avoided. The interference pattern at the liquid surface is stationary for continuous oscillations. A hologram can be recorded on a film, or it can be projected in a straightforward way optically; a laser light beam, L, is transmitted through a semi-transparent mirror, Z, on a holographic plane, is reflected through an optical system and is projected on a screen where a reconstructed picture of the object can be seen. The aperture, C, serves as a spatial filter for the elimination of multiple images.

This principle is primarily of significance in the laboratory, because the transmission method is unsuitable for the defectoscopy (NDT) in practice. The reflection pulse method seems to be more suitable for testing of materials. The circuit diagram is indicated in Fig. 5.4. The examined object, 1, is immersed in a bath, 2, filled with a liquid, 3. A small probe, 4, above the object, 1, moves in strokes (scanning lines) and it alternately emits and receives ultrasonic pulses. The pulses are relatively long so that the duration of the stationary state lasts relatively longer than the transient state at the start and at the end of the pulse. The probe, 4, is led by a scanning support, 5.

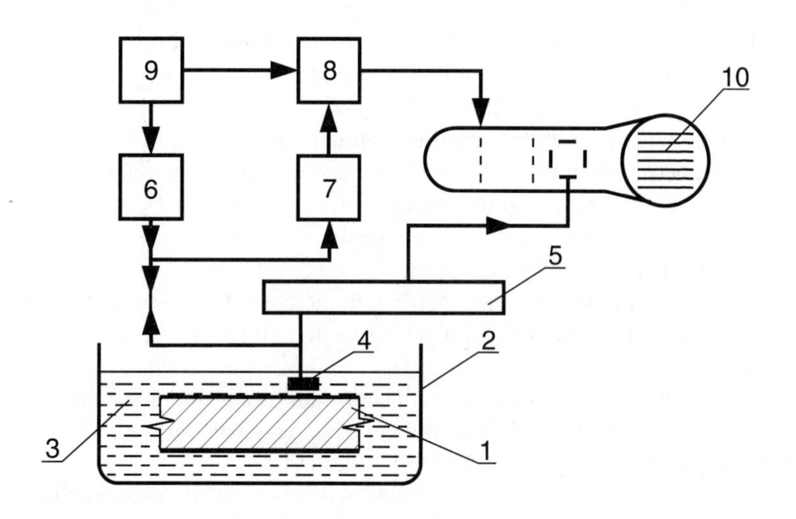

Fig. 5.4 ***Schematic diagram of an instrument for pulse reflection holography***
1 - object, 2 –bath, 3 - liquid, 4 - ultrasonic probe, 5 - mechanical scanning system, 6 - ultrasonic pulse generator, 7 - amplifier, 8 - mixer, 9 - source of reference frequency, 10 - CRT

The liquid mediates as an acoustic coupling material. The ultrasonic probe, 4, is supplied, and excited, from a generator, 6, by rectangular pulses of high frequency, from a source generator at the reference frequency, 9. The pulses surfacing from a certain depth of the examined object are screened, amplified, and then they pass to a mixer, 8, simultaneously with the reference frequency from the frequency source, 9. In the mixer, the interference is created electronically. The output voltage of the mixer, 8, modulates the brightness of a CRT screen, 10, and in this way, the hologram is visualized on the screen. An advantage is that access from only one side of the examined object is sufficient. A reconstruction of a hologram is illustrated in Fig. 5.5. A light beam from a laser, 1, with a wavelength of about 0.5 μm, is scattered in the lens, 2, and passes through an additional optical system, 3, into which a hologram, 4, is inserted, usually as a developed photographic film. The beams for which no diffraction was captured in the hologram are rejected in the aperture, 5. Through a further lens, 6, the actual hologram is projected on a screen, 7.

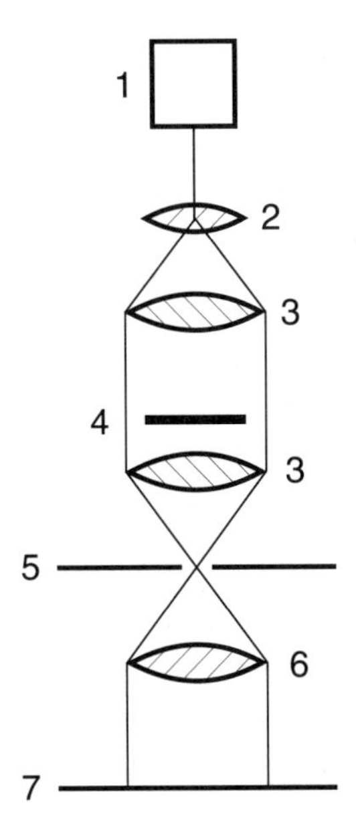

Fig. 5.5 ***Design of an arrangement for reconstruction of a hologram***
1 - laser, 2 - lens, 3 - optical system, 4 - hologram, 5 - aperture, 6 - lens, 7 - screen

5.2.6 The acoustic emission method

This method of testing of materials, which utilizes acoustic emission, is based on analysis of the signals which originate during the broadening of a rupture in a material. Acoustic emission arises as a result of the release of energy of transition from the elastic to the plastic deformation state of a material. The stored deformation energy is released at the point of the initial disturbance of the original structure of a material. Part of this released energy is transformed to sonic and ultrasonic oscillations in the form of pulses, whose spectral band extends from kHz up to some MHz. Physically the origin of the pulses of acoustic emission can be explained by two reasons. The first of these is a mutual shift of dislocations which accompany plastic deformation. A dislocation shift is a source of an acoustic pulse, which propagates isotropically through a medium, as the dimensions of the source are practically point-like. A simultaneous movement of tens or hundreds of dislocations results in an acoustic pulse which can be detected by an ultrasound receiver. The second reason is due to the effects accompanying ruptures or fissures in a material, and as a rule they are more intensive acoustic sources than those from the shift of dislocations. Thus microcracks arise and their spreading is accompanied by a release of energy resulting from a rapid decrease in mechanical tension. On releasing the tension, the acoustic emission signals disappear, and if loading is renewed, they appear again only above the last (highest) level of tension (the so called Kaiser effect).

It follows from the foregoing that an acoustic emission can be continuous or pulse-like. A continuous emission is obtained at a transfer of dislocations. The formation of ruptures, or their propagation, yields pulse signals. For testing of materials, only pulse emission is important. Its advantage is that its intensity is higher by up to two orders of magnitude than that of continuous emission.

The acoustic emission signals are picked up by piezoceramic probes, adapted for receiving surface waves. Their maximum sensitivity is always only within a certain limited frequency band, in order to eliminate the possibility of receiving unwanted interference which arises from the operation of the equipment.

The acoustic emission signals are processed by their number, and by their intensity of amplitude. The simplest way is to count the number of overshoots above a certain threshold value, U_p (Fig. 5.6). An emission event (Fig. 5.6a) can be characterized by the amplitude and frequency of the ringing.

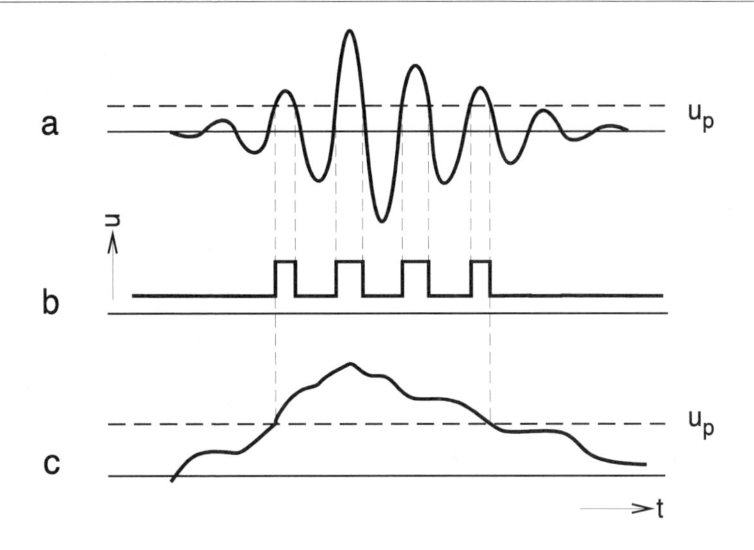

Fig. 5.6 ***Emission event***
a - shape of a signal received by a probe, b - shaped overshoots, c - an envelope curve of the emission event after rectifying and filtering

The threshold level, U_p, can be preset. In this way, a level can be preset on crossing of which the overshot signal is shaped into a rectangular pulse. These pulses (Fig. 5.6b) can be recorded by a counter. A single emission event may be identified by several overshoots. If it is desirable to count a number of emission events only, the best way is to rectify and filter the hf pulses corresponding to the individual emission events, or to count in only the first overshoots. Through rectification, an envelope curve of the emission event is created. The filter time constant is preset according to the selected frequency band, to the number of overshoots and, in particular, according to the repetition frequency of the individual emission events.

The acoustic emission can also be processed by an amplitude distribution analysis (pulse height analysis, PHA mode), or if need be, by spectral analysis as well.

The block circuit diagram of an instrument for receiving acoustic emission signals is shown in Fig. 5.7. A signal from the probe, 1, passes through a preamplifier, 2, to an amplifier, 3, which, if necessary, can be equipped with a filter. Several different circuits can be connected at its output. The discriminator of overshoots, 4, is connected with a counter, 5, a filter for the envelope curve, 6, of the emission event and a counter of events, 7. Normalized amplifiers, 8 and 10, follow, to which the amplitude analyser, 9, and the frequency analyzer, 11, are connected. The results can be recorded either on an X-Y recorder, 12, or on a digital recorder, 13. When the data are processed by an on-line computer, the output is connected to a computer interface, 14 [5.3].

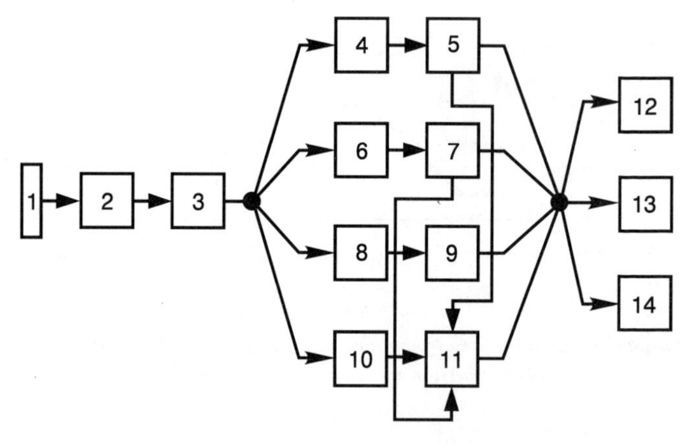

Fig. 5.7 ***A block diagram of processing signals of an acoustic emission***
1 - probe, 2 - preamplifier, 3 - electrical filter with an amplifier, 4 - discriminator, 5 - overshoot counter, 6 - filter for the envelope curve of the emission event, 7 - counter of emission events, 8 - amplifier, 9 - amplitude discriminator, 10 - amplifier, 11 - frequency analyzer, 12 - recorder, 13 - plotter, 14 - computer interface

5.3 Special methods

5.3.1 Imaging of defectoscopic signals

Signals obtained from the probe of an ultrasonic system by transition through a medium or by a reflection of the ultrasonic pulse can be processed and projected in several ways. The so called projection A, described in section 5.2.2, is considered classical. Besides this, further projections called B and C are known.

In projection A, the time base is displayed on the horizontal scale of the screen and the height of the received echo on the vertical scale. Usually a CRT screen is used for this projection. In this case the time scale is calibrated in length units.

In projection B, both the coordinates correspond to length. The vertical coordinate usually represents the depth of a defect, and the horizontal one the position of the moving probe along the surface of the object (Fig. 5.8). The position of the defect, 3, is electronically transferred to the vertical deflection plates of a CRT. The sawtooth time base voltage is transferred to the horizontal deflection plates, similarly to projection A. The echo is represented by a modulation of the intensity of the screen brightness. So in fact, a particular cross section of the object is projected. With regard to the slow movement of the probe along the object surface, a storage CRT must be used in order to retain the image, or alternatively the recording is digitized and the projection is carried out after the reconstruction of the recording. The digitization and the subsequent processing of the image by computer enables one to carry out suitable filtering of the signal and calculation of the position of a defect and of its dimensions.

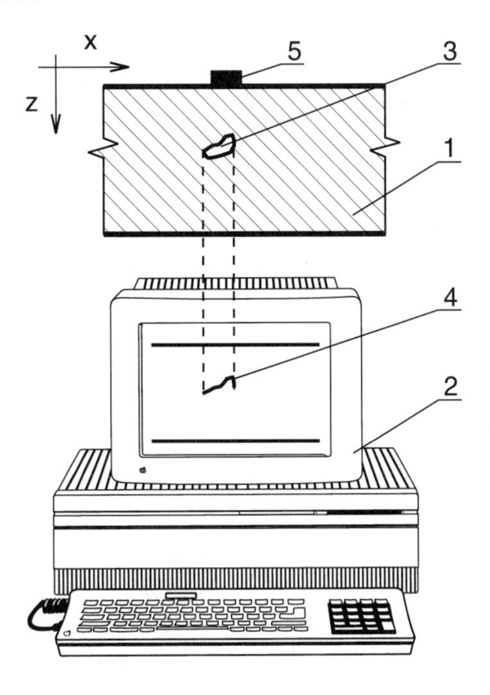

Fig. 5.8 ***Projection B***
1 - medium, 2 - CRT screen, 3 - defect, 4 - defect projection, 5 - ultrasonic probe

Projection C allows scanning of defects in a plane, using X-Y coordinates. In these coordinates, the probe moves above the surface of the object, and it scans the whole surface area. Besides projection in X-Y coordinates, planar projection can be performed also as a projection in polar coordinates. For this reason, a CRT is used whose beam rotates in a synchronous way with the probe movement. This arrangement has proven successful especially when testing metal sheets by using surface waves.

5.3.2 Spectral ultrasonic method

The methods described above are perfectly sufficient in the cases where only the determination of the size and the depth of a defect are necessary. However, it is often important also to obtain information about the shape, and of the orientation of the defect. Therefore alternative arrangements were investigated for extracting new information from the ultrasonic signal, which may be contained, for instance, in the shape of the defect, on which the frequency spectrum depends. One way of employing spectrometry in ultrasonic defectoscopy is an examination of the changes produced in the ultrasonic pulse spectrum after transition through the medium under investigation.

The spectrum of a pulse can be obtained by Fourier analysis by which the shape of a non-periodic pulse (5.4) can be described by the function

$$f(t) = \frac{1}{2\pi} \int_{-\infty}^{\infty} S(\omega)\, e^{j\omega t} d\omega \qquad (5.4)$$

where $S(\omega)$ is the complex spectrum of a particular non-periodic function $f(t)$.

The modulus of the complex spectrum, called simply the spectrum, is identical with the absolute value of the Fourier image, $F(\omega)$,

$$S = |S(\omega)| = |F(\omega)| = \left| \int_{-\infty}^{\infty} f(t) e^{-j\omega t} dt \right| \qquad (5.5).$$

It describes the dependence of the individual terms on frequency. The continuous spectrum given by the expression (5.5) is dependent on the function f(t), which represents the shape of the pulse and creates an envelope of the amplitudes in the discrete line spectrum of a periodic pulse. Thus, the shape of the spectrum is not dependent on the repetition frequency. The relationship of the amplitude spectrum, $S(\omega) = S$ of a high frequency pulse, and its envelope curve after detection (a videopulse), is an example, shown in Fig. 5.9. The shapes of the spectra S are similar, with the only difference being that the maximum ampli-

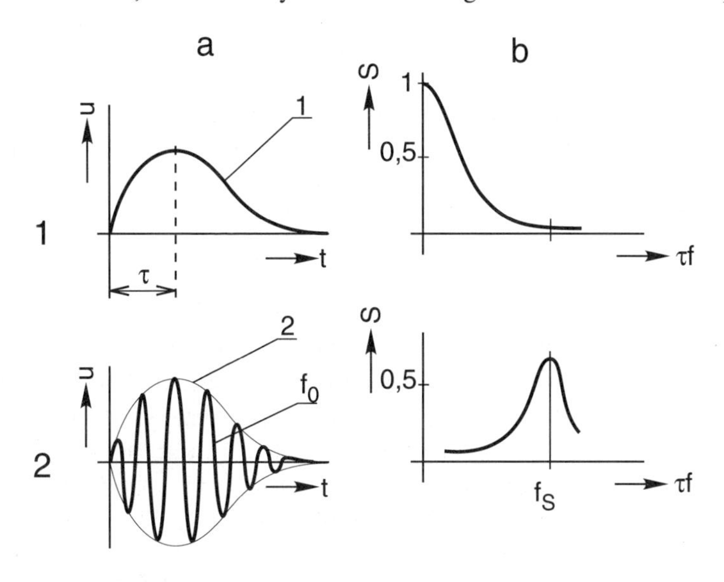

Fig. 5.9 ***Spectrum of an ultrasonic pulse***
1 - videopulse, 2 - HF pulse: a) - time mode; b) - frequency mode

tude of the spectrum of a high frequency pulse has a value of one half of the value for a videopulse spectrum. It is caused by a symmetrical distribution of the high frequency pulse spectrum around the mean frequency, f_S. The spectrum of a videopulse begins with a dc term. As a mean frequency, the frequency term with the maximum amplitude is denoted which corresponds to the frequency of the ultrasonic oscillations. An ultrasonic pulse in acoustic media, including the electroacoustical transducer, the probe and the electronic circuits, can be analyzed with advantage in the frequency mode, i.e., as a spectrum. The properties of a circuit are described by a so called response, i.e. by a signal u(t) at the circuit output, induced by an input excitation, f(t). The frequency response is determined by the frequency characteristics, or by the transient function, $H(\omega)$. If a signal f(t) is presented to the input of a linear system with a pulse response h(t), the resultant response in the time mode is given by the convolution

$$u(t) = f(t) * h(t) = \int_{-\infty}^{\infty} f(t)\, h(t-\tau)d\tau \qquad (5.6).$$

The frequency response can be determined by multiplying the input signal (or rather, its spectral shape) in the frequency mode, $F(\omega)$, by the transfer function $H(\omega)$

$$U(\omega) = H(\omega) * F(\omega)\ .$$

Ultrasonic spectrometry is based on an evaluation of the response $U(\omega)$, in which the properties of the examined material are represented by the functions $D(\omega)$ and $R(\omega)$. Therefore the acoustic medium and the character of the defects can be described by spectrometry, which processes the amplitude spectra simultaneously with the phase spectrum.

The equipment required for frequency analysis of a spectrum consists of classical ultrasonic pulse instruments fitted with wide-band probes. A selected pulse or a series of pulses are introduced to a spectrum analyzer, either in an analogue or a digital form. In the second case, both the amplitude and the phase spectra can be obtained, while an analogue analyser evaluates only an amplitude spectrum.

Besides inspection of the character and orientation of defects by means of ultrasonic spectrometry, other parameters of a medium can be determined, such as its attenuation, structure and so on.

5.4 Design and construction of NDT instrumentation

Ultrasonic instruments for non-destructive testing, NDT, consist of three fundamental parts:

a - probe
b - electronic circuits for processing the echo
c - display projection.

5.4.1 Ultrasonic probes

The task of ultrasonic probes is to transmit and to receive ultrasonic pulses. According to their purpose, we can classify them as

- direct probes
- angular probes
- probes which focus the beam
- pick-ups of acoustic emission.

Apart from this classification scheme, further alternatives and designs of probes for special requirements and purposes are known.

Direct probes
Direct probes transmit and receive ultrasonic pulses perpendicularly to their contact faces. They can operate with a longitudinal or with a transverse wave. The simple direct probes are equipped with a single piezoelectric transducer, which transmits as well as receives ultrasonic waves. The most important parameters of a probe are sensitivity, good resolution and a small dead zone. The necessary bandwidth is obtained by using a suitable piezoceramic transducer and, above all, by damping its rear side. From one side (Fig. 5.10), the piezoceramic transducer, 2, is separated from the medium by a thin protective layer, 1. The damper body, 3, is suitably shaped in order to behave as an infinite medium from which no pulses are reflected back to the transducer [5.5]. The front protective and covering layer is usually tuned via its thickness to a quarter-wave resonance:

$$dv = (2n + 1)\frac{\lambda v}{4} \quad \text{for } n = 0, 1, 2, \ldots$$

Thus, the widest possible band is obtained, and consequently the highest resolution.

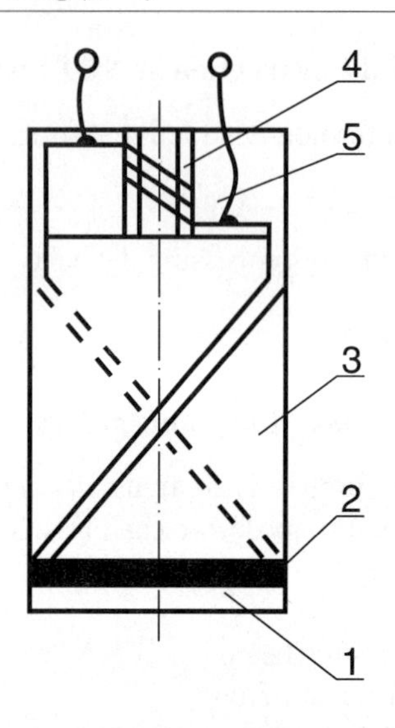

Fig. 5.10 ***Pulse probe***
1 - covering (separating) layer, 2 - piezoelectric transducer, 3 - damper with directional grooves, 4 - matching coil, 5 - connecting leads

The mechanical and the electrical damping of the probe affect the sensitivity as well as the axial resolution of the probe. According to the damping, the probes are classified as undamped (narrow bandwidth), lightly damped (standard) and heavily damped (broad bandwidth).

The narrow bandwidth probes ($\Delta f/f_s < 0.4$) have a high sensitivity and a low resolution. The standard probes ($\Delta f/f_s < 0.8$) are less sensitive and their resolution suits the common applications of defectoscopy. The broad bandwidth probes ($\Delta f/f_s \cong 1$) have small sensitivity and high resolution.

The narrow band probes and the standard ones have an electrical resonant circuit connected to the piezoceramic transducer. The resonant circuit is created by an inductance, L, and a capacitance, C, a part of which is due to a contribution from the transducer capacitance and the leading cable capacitance. The resonant circuit is shorted by a variable damping resistor, R, connected in parallel, by which the bandwidth can be adjusted.

For spectrometry, wide-band probes are required, with the ratio ($\Delta/f_s \geq 1$) (Fig. 5.11c). These probes must be constructed from special piezoceramic transducer materials such as lead metaniobate, which exhibits very narrow planar oscillations. It permits one to connect the transducer to the generator and the amplifier without an electrically tuned circuit. This circuit, in connection with narrow bandwidth probes, rejects all frequencies except the narrow working band.

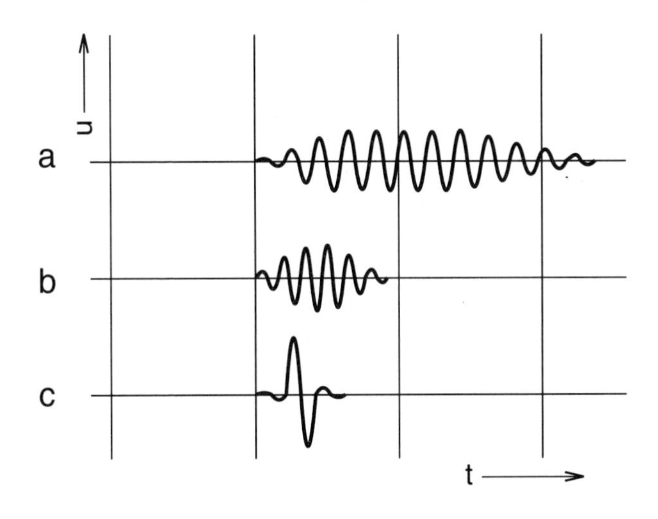

Fig. 5.11 ***Shape of an ultrasonic echo from the probe a - undamped, b - lightly damped, c - wide-band***

The direct ultrasonic probes used for defectoscopy have transducers with effective diameters, D_{ef} = 5-30 mm. It is recommendable to use a radiation pattern with a half angle of the major radiation lobe of up to 10°. The probes are protected against the influence of external electric fields by a metal casing. The outer (surface) electrode of the piezoelectric transducer is grounded.

If a transversely polarized transducer is used inside the probe, then a simple direct probe for transverse waves is obtained.

If identifying defects in shallow depths, for instance in cases when simple probes cannot be used because of their dead zones, double probes are recommended. These employ one transducer for emission, and another one for reception. The double probes exhibit the highest sensitivity at a given depth below the surface. This depth depends on the tilt angle of the transducers. An advantage of using the double probes is an improved signal-to-noise ratio.

Angular probes

Angular probes emit and receive ultrasonic waves at a particular angle relative to the medium surface and they are based on the principle of refraction of ultrasonic waves (see Chapter 2). They serve for testing materials using transverse as well as longitudinal waves. The interface at which the refraction arises is created by a so called refractory wedge. On the surface of the tested material, a longitudinal ultrasonic wave is incident at an angle α_1 and it is refracted at an angle α_2. The refractory wedge is made from a material whose velocity of wave propagation is lower than the velocity of the same kind of waves in the test medium. The most commonly used material is plexiglass. In order that the unwanted waves do not return back to the transducer after reflection, they are attenuated in the refractory wedge, 4 (Fig. 5.12). The transducer, 1, is damped by a damper body, 3, similarly to the direct probes.

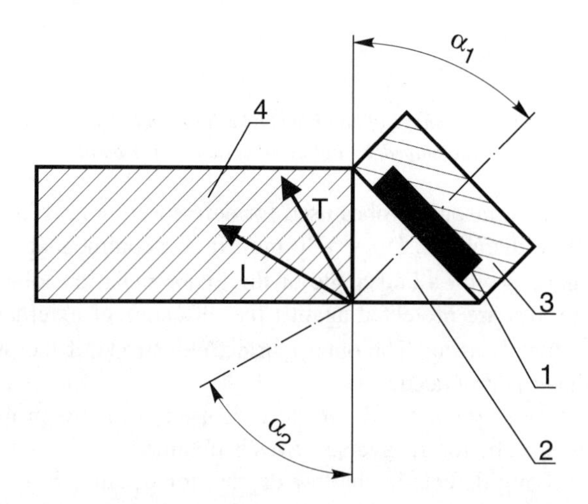

Fig. 5.12 ***Angular probe with a damping medium***
1 - transducer, 2 - refractory wedge, 3 - damper body, 4 - damping medium of the refractory wedge

Focused probes

The main task of focused probes is to concentrate the ultrasonic energy into a small space, in order to increase the sensitivity, resolution, precision of determination of the defect position and the signal-to-noise ratio. The most common and the simplest means of focusing is a lens. Due to refraction, the ultrasound concentrates to a focal point (Fig. 5.13a). Similarly, ultrasonic energy can be concentrated by shaping the ultrasonic transducer to a spherical shape (Fig. 5.13b).

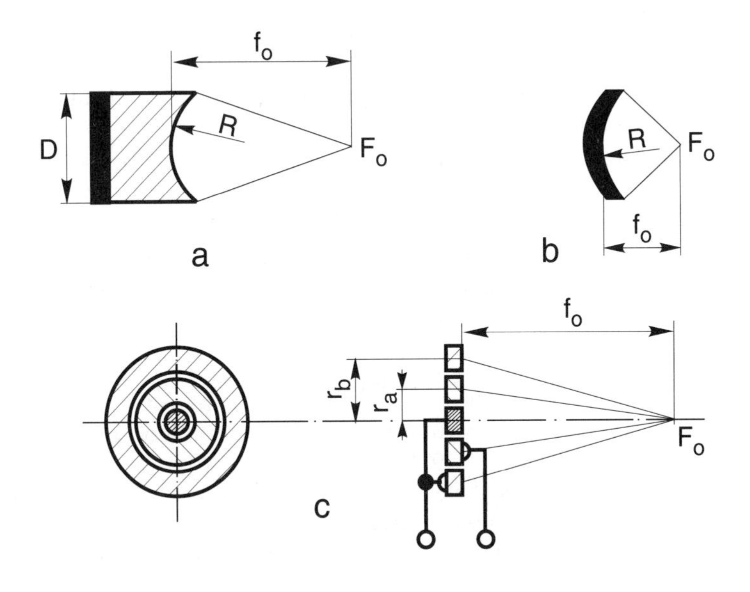

Fig. 5.13 ***Focusing of ultrasonic waves***
a) - planar transducer and an acoustic lens, b) - curved transducer,
c) - a set of time-phased transducers

An ultrasonic beam can also be focused by a phase-dependent (phase-timed) set of transducers, shaped as coaxial annular rings and arranged around the central round transducer (Fig. 5.13c). To the electrodes of the even and the odd annular rings, voltages of alternate polarity are applied. Therefore this method of exciting transducers is also called phase focusing.

Receivers of acoustic emission

The signals from acoustic emission received by pick-ups laid on the surface of the material are usually surface waves. For this reason, they cannot be detected by probes, which are adapted for longitudinal or transverse ultrasonic waves. The pick-ups for acoustic emission utilize planar oscillations, and therefore they are usually tuned to a resonance frequency which corresponds to the radial dimensions of the transducer. The resonant frequency of such a transducer is

$$f_r \approx \frac{c_r}{2D} \tag{5.7}$$

where c_r - velocity of propagation of the surface waves inside the transducer

D - transducer diameter.

The frequency characteristics of such pick-ups of acoustic emission are mostly wide-band ones, because the spectrum of a signal is, as a rule, not known beforehand. The bandwidth is usually from 200 to 300 kHz.

5.5 References

5.1 *Metherell A.F. et al:* Acoustical Holography, Plenum, New York Vol. 1: 1969, Vol. 2: 1970, Vol. 3: 1971

5.2 *Obraz J.:* Ultrazvuková holografie a její použití, Jemná mechanika a optika, 16, 1973, No. 4, 94-96 (in Czech)[1]

5.3 *Dunengan H.L.,Tetelmann A.S.:* Acoustic emission, Research Development, 22, 1981, No. 5, 20-24

5.4 *Fitting D.W., Adler L.:* Ultrasonic Spectral Analyses for Non-Destructive Evaluation, Plenum, New York, 1981

5.5 *Figura Z. et al:* Tlmiace teleso ultrazvukovej impulznej sondy, ČSSR Author's Certificate No. 265541, from 10.2.1989 (in Slovak)[2]

[1]Ultrasonic holography and its utilization, Fine mechanics and optics

[2]A damping body of an ultrasonic pulse probe

6 Measurement of position and air flow by pulse methods

The contents of this chapter resume the discussion from section 4.3. It represents an enhancement, and appends one significant solution which has been developed in greater detail by research. The method described next enables more efficient rejection, or compensation, of all atmospheric influences. For compensation of the effect of air flow, the method requires introducing a further measurement. However, for carrying out this measurement, the same measuring apparatus is used repeatedly, with merely an extra switch of two reciprocal transducers attached, whereby the transducers, in their basic measuring mode, can be single-functioned. Other operations are performed by software. The software is enhanced by a program which calculates the air flow vector terms, and a subsequent correction to the coordinates calculated originally. This method is also significant from the point of view of an independent measurement of air flow, such as the wind. Therefore a separate section is devoted to this.

6.1 A pulse method of coordinate measurement using a reference transducer, positioned at the coordinate origin

Using an auxiliary reference transducer, positioned at the origin of a coordinate system, represents a significant increase in the precision of a measurement. It allows measurement of one extra distance, d_o, and thus introduces one more equation to the original system of equations. This enables one to calculate the value of the coefficient, k_t, which was the main source of errors. This coefficient represented the influence of atmospheric parameter fluctuations in the calculation and caused a significant error even for small variations of these parameters.

6.1.1 Measurement of coordinates using point microphones

A configuration of ultrasonic transducers utilizing the method described above is illustrated in Fig. 6.1. According to the geometry of the trans-

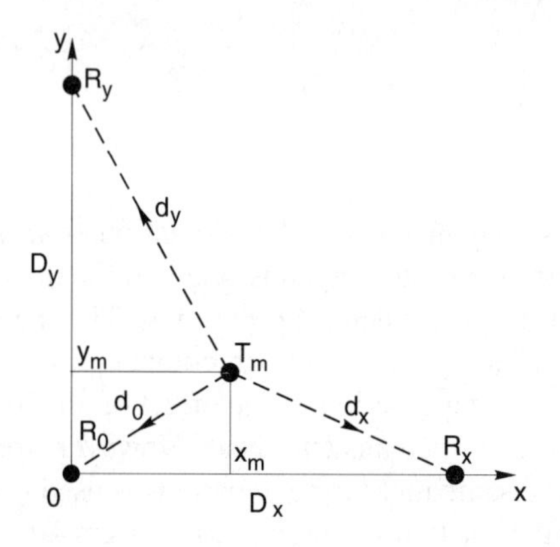

Fig. 6.1 ***Arrangement of ultrasonic point transducers for measuring planar coordinates, using a reference transducer positioned at the origin of the coordinates***

ducer arrangement, a system of equations can be written:

$$y_m^2 + (D_x - x_m)^2 = k_t^2 N_x^2$$
$$x_m^2 - (D_y - y_m)^2 = k_t^2 N_y^2$$
$$x_m^2 + y_m^2 = k_t^2 N_o^2 \qquad (6.1).$$

The third equation constrains the first two and enables one to calculate the value of the coefficient, k_t, assuming a uniform distribution of air parameters in the measured plane. Thus the isotropic propagation of an ultrasonic wave (in all directions) is assumed. The influence of the air flow is therefore not compensated.

By solving the system of equations, a biquadratic (quartic) equation is obtained

$$a_4 k_t^4 + a_2 k_t^2 + a_0 = 0 \,,$$

and its solution yields a value k_t^2. The individual coefficients a_i are given by expressions

$$a_4 = D_x^2(N_o^2 - N_y^2)^2 + D_y^2(N_o^2 - N_x^2)^2$$
$$a_2 = -2D_x^2D_y^2(N_x^2 + N_y^2)$$
$$a_o = D_x^2D_y^2(D_x^2 + D_y^2)\ .$$

The subsequent solution of the system of equations (6.1) yields the calculation formulae for the coordinates,

$$x_m = \frac{1}{2D_x}\left[D_x^2 + k_t^2\left(N_0^2 - N_x^2\right)\right]$$
$$y_m = \frac{1}{2D_y}\left[D_y^2 + k_t^2\left(N_0^2 - N_y^2\right)\right] \qquad (6.2).$$

Before calculation of a coordinate, the coefficient k_t^2 is first calculated. For a known and stable value of frequency f the velocity c can be calculated as well. In contrast to the measurement without a reference transducer, this time we are able to determine the coordinates of the measured point unambiguously. This results from the assumption that in a plane, a single point corresponds to a single ratio of distances $d_x : d_y : d_o$ (or, if need be, $N_x : N_y : N_o$).

Miniature microphones are a good approximation to point-like ones in the case of ascertaining large distances compared with the size of their active surface. A similar case is the use of a miniature spark gap as a source of an ultrasonic wave pulse. This was treated in more detail in Chapter 3.

6.1.2 Measurement of coordinates using cylindrical transducers

The precision of a measurement can be increased by using transducers with a cylindrical active surface. These cylinder-transducers are mostly reciprocal, and they can therefore also be used for generating an ultrasonic wave pulse. Point-like microphones cannot fulfil this task because of their small power. Such a case is indicated in Fig. 6.2, with cylinder-shaped transducers with equal radius, r, of their active surface area. Even if the entire cylindrical surface is not exploited, they are as easy to manufacture as the capacitive or the piezoelectric types of transducer.

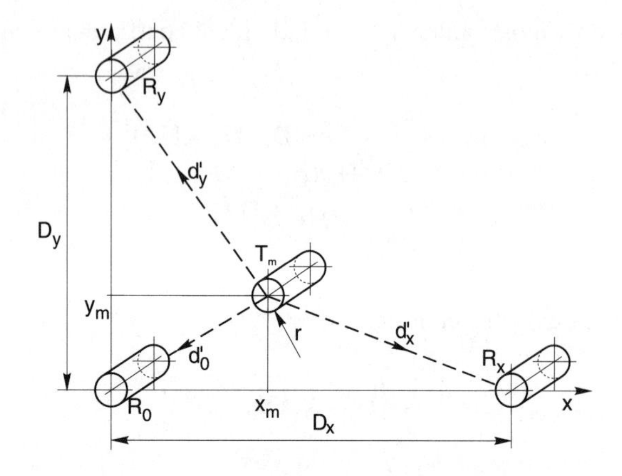

Fig. 6.2 ***Arrangement of cylindrical transducers for measuring planar coordinates using a reference transducer positioned at the coordinate origin***

However, the measurement can in general be performed as a four-quadrant one.

Even from cursory inspection of Fig. 6.2, it is clear that the original distances d_x, d_y, d_o are, at the unchanged position of the transducers, shorter by a length equal to two radii, 2r. This results in smaller measured values N'_x, N'_y and N'_o. It leads to a more complicated calculation of the coefficient k_t, which can be calculated (for this particular case) from the 4th order equation [6.4]:

$$a_4 k_t^4 + a_3 k_t^3 + a_2 k_t^2 + a_1 k_t = 0 \,.$$

The expressions for calculation of the coefficients a_i are relatively complex. However, given up-to-date computational facilities, the calculation can be performed in such a short time that the measurement is carried out practically in real time. With smaller variations of the parameters, especially temperature, a simplification can be introduced by means of which the 4th order equation introduced above transforms to a biquadratic one. If a particular propagation velocity, c_o, of an ultrasonic wave is assumed, at the average (ambient) working conditions (e.g. 20 °C), then a number N_r can be expressed which would be measured during propagation of the pulse from the active transducer surface to its center, or alternatively in the opposite direction,

$$N_r = f_r/c_o\,.$$

The number N_r should be added to the measured values (twice since both transducers are cylindrical), and in this way we arrive at the numbers which are to be substituted for the numbers on the right hand sides of the equations (6.1)

$$N_x = N'_x + 2\,N_r\,; \quad N_y = N'_y + 2N_r; \quad N_o = N'_o + N_r.$$

The measurement errors, as a result of this substitution, will increase in proportion to the deviations of the actual parameters from the assumed ones.

6.1.3 Measurement of spatial coordinates

As in the case of measurement of planar coordinates, the system of transducers in Fig. 6.4 can be augmented by a reference transducer, R_o, placed at the origin of the coordinate system (Fig. 6.3). Relations for

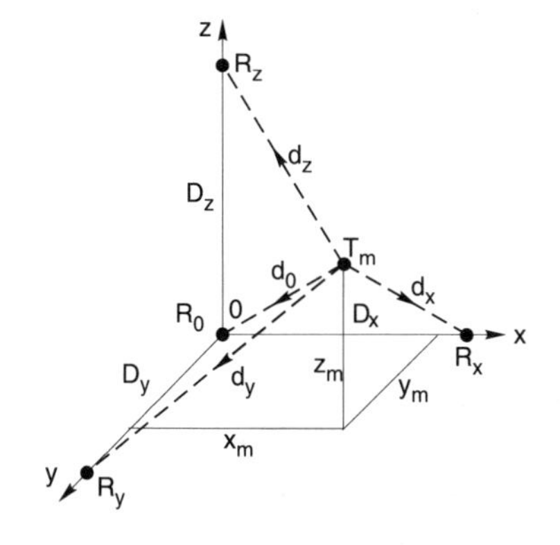

Fig. 6.3 ***Arrangement of point transducers for measuring spatial coordinates using a reference transducer at the coordinate origin***

calculation of the spatial coordinates can be derived by a procedure like the one described earlier in the case of planar coordinates for point-like transducers (§ 6.1.1), as

$$x_m = D_x/2 + k_t^2(N_o^2 - N_x^2)/2D_x$$
$$y_m = D_y/2 + k_t^2(N_o^2 - N_y^2)/2D_y$$
$$z_m = D_z/2 + k_t^2(N_o^2 - N_z^2)/2D_z \qquad (6.3).$$

Again, the relations for calculation of the coefficient k_t^2 are obtained by solving the biquadratic equation, in which the following values are substituted for the individual coefficients,

$$a_o = D_x^2 D_y^2 D_z^2 (D_x^2 + D_y^2 + D_z^2)$$
$$a_2 = 2D_x^2 D_y^2 D_z^2 (N_o^2 - N_x^2 - N_y^2 - N_z^2)$$
$$a_4 = D_y^2 D_z^2 (N_o^2 - N_x^2)^2 + D_x^2 D_z^2 (N_o^2\ N_y^2)^2 + D_x^2 D_y^2 (N_o^2 - N_z^2)^2.$$

In the case of substitution of point-like transducers by spherical ones with a radius r, the need for the solution of a 4th order equation arises again. This equation, however, can be transformed in the manner described above to a biquadratic one.

Compensation of the influence of parameter variations in a uniform (i.e. non-gradient) medium (except for compensation of the influence of the air flow) is achieved again by introducing a reference transducer, with an appropriate electronic circuit. As the air flow is expressed by a vector field and its compensation is more demanding, a separate section is devoted to this problem.

6.2 Measuring air flow

Ultrasonic measurements require a medium in which the propagation of ultrasonic waves is physically possible. If the measurements are performed in a gaseous or liquid medium, then motion of the medium can occur during the measurement. The flow velocity of the medium is a vector field, and it is also measurable by means of ultrasound [6.5].

6.2.1 Principle of measurement

Let us suppose a measurement is carried out in flowing air which has a uniform velocity throughout the whole measurement volume. Let the air flow be homogenous. The ultrasonic wave pulse is carried along by the flowing medium with a velocity v. Let us further suppose the ultrasound propagation velocity be known, and has a value c_o. On measuring the medium flow velocity v, which is a vector quantity, we need to separate the term c_o, because the measured (resultant) velocity is composed of both velocities, as their vectorial sum [6.14].

Measurement of the flow velocity is based on a simple consideration. Firstly, the time of propagation of an ultrasonic pulse is measured from a transmitter, T_1, to a receiver, R_l (Fig. 6.4). Then the positions of

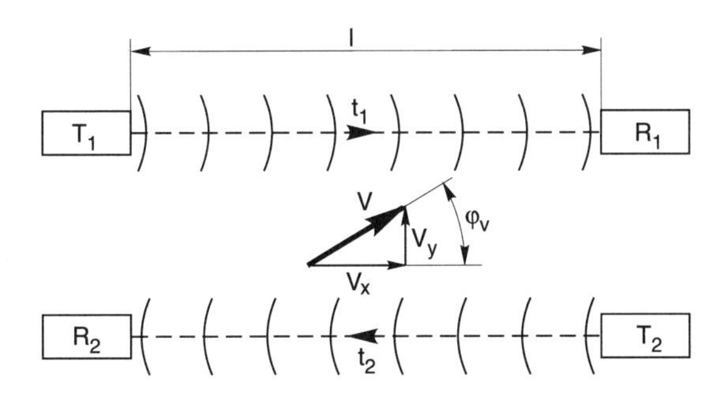

Fig. 6.4 ***Illustration of the principle of measurement of flow velocity***

the transmitter and the receiver are reversed and the time of propagation, t_2, is measured again in the opposite direction, from T_2, to R_2. If reciprocal transducers are used, then they remain in the same positions as in the first measurement, but their functions are merely reversed. The time difference $\Delta t = t_2 - t_1$ gives information about the air flow velocity in the direction of the line connecting transmitter to receiver [6.14],

$$\Delta t = t_2 - t_1 = \frac{1}{c_0 - v_x} - \frac{1}{c_0 + v_x} = \frac{2 l v_x}{c_0^2 - v_x^2} \cong \frac{2 l v_x}{c_0^2} \quad (6.4).$$

In the approximate expression on the right hand side, the presumption is made that $c_0 \gg v_x$. From (6.4), we obtain a relation for the flow velocity, v_x,

$$v_x = \frac{c_0^2}{2l}\Delta t = k\Delta t \qquad (6.5).$$

These expressions hold true if the direction of the flow is parallel with the connecting line from transmitter to receiver. If the air flows at an angle φ_v to this connecting line, the result of the measurement is the cosine of the flow velocity vector (Fig. 6.4). The measurement principle, according to Fig. 6.4, is applied in practice using one-dimensional apparatus - flow gauges. In these cases, the required direction of the medium flow is ensured by their design and construction.

In the technical realization of this measurement principle, the need arises to use reciprocal pulse transducers. The capacitor and the piezoelectric transducers suit this demand. However, their realization is more demanding than that of plain, single-function transducers, and therefore as an alternative, pairs of single-function transducers are also used, placed close to each other.

A considerable advantage of ultrasonic measuring methods is that the spatial propagation of ultrasound allows a relatively simple switch from the one-dimensional (1D) measurements to planar (2D) ones, and also to spatial (3D) ones. To accomplish this task, an increase in the number of transmitters as well as of receivers is necessary, as well as their consequent suitable fastening in a plane or in space.

6.2.2 Measurement of a planar vector of the air flow

In Fig. 6.5, three ways of measuring a planar air flow vector are indicated.

In the measurement shown in Fig. 6.5a, independent measurements of the vector terms, v_x and, v_y are carried out simultaneously. The direction and the velocity of the air flow are calculated from:

$$\varphi_v = \arctan\frac{v_y}{v_x} \qquad (6.6)$$

$$v = v_x^2 + v_y^2 \qquad (6.7)$$

whereby the individual terms are calculated according to (6.5). The appropriate quadrant is identified according to the signs of the terms v_x and v_y.

In the measurement shown in Fig. 6.5b, one reciprocal ultrasonic transducer is saved, at a cost of splitting the measurement into a se-

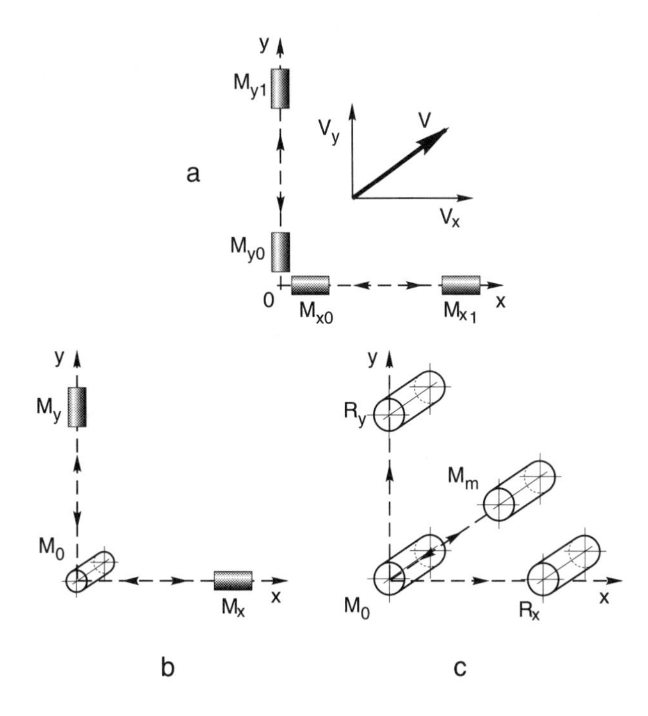

Fig. 6.5 ***Transducers and their configuration for measuring a planar (2D) vector of the flow velocity***

quence of two phases, and needing to use a reciprocal transducer, M_o, with a wide radiation pattern. In the measurement shown in Fig. 6.5c, a method of measuring planar coordinates, originally presented in section 6.1.2 (Fig. 6.2), is recapitulated. The only difference is that the transducers T_m and R_o are exchanged for reciprocal ones, M_m and M_o, so that the second phase of the flow measurement can be carried out [6.11, 6.6]. In this latter phase their function is exchanged. Only a single transducer switch needs to be added to the electronic circuit. The original electronic circuits remain unchanged. The main change, however, is in the control program and processing of the measured data by the computer, by the software.

6.2.3 Measurement of a spatial vector of the flow velocity

In the same way in which the 1D measurement was extended to the 2D measurement, it is possible to change from 2D to the 3D measurement of

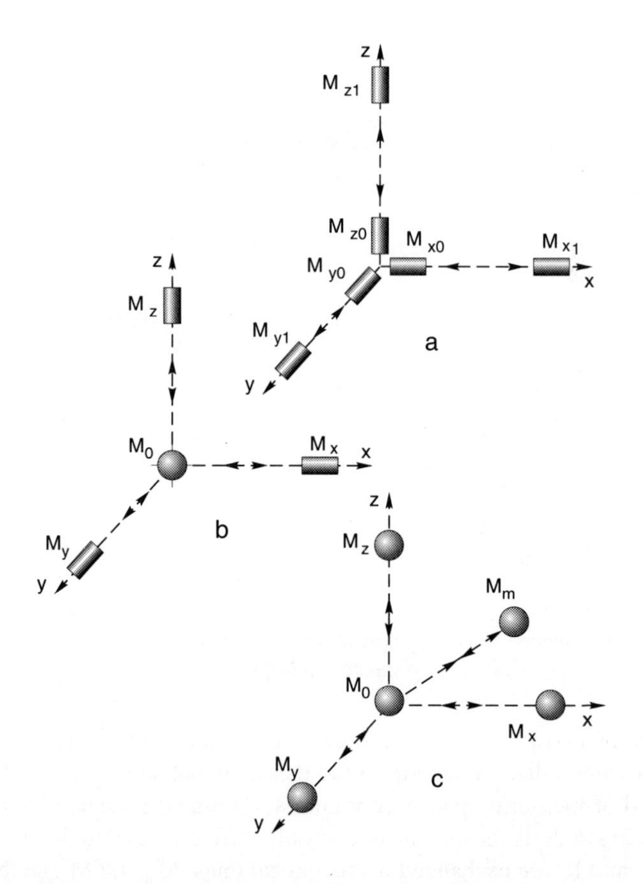

Fig. 6.6 ***Transducers and their configuration for measuring a spatial (3D) vector of the flow velocity***

a spatial vector of the flow velocity. In essence, the procedure rests on extending the measuring equipment by adding a further measuring channel, positioned on the z axis (Fig. 6.6). As in the case shown in Fig. 6.5a,b, it is possible to evaluate the direction and the velocity terms, v_x, v_y and v_z, of the medium flow from the measured velocity terms (Fig. 6.6).

6.3 Measurement of the position and flow with a single instrument

The method of measuring the flow indicated in Fig. 6.5.c was introduced deliberately in order to illustrate correction of the coordinates using a single device. Since the method of coordinate measurement has already been described, in the next section we will focus our attention on flow measurement alone, and on the correction of coordinates. However, the method can be used separately for independent measurement of the air flow, for instance for measuring the wind in meteorology.

6.3.1 Measuring flow in a plane

The essence of the method and the derivation of the relevant expressions can be explained more usefully by examining the case of measurement of a planar flow. For spatial measurement, the expressions can be extended straightforwardly. The reverse is also possible, though. The relations for spatial measurement can be considered to be complete, and the expressions for planar measurement are obtained simply by elimination of variables related to the third dimension.

In order to simplify the relations, point-like transducers are again assumed. In Fig. 6.7, the spatial arrangement of the transducers and their distances is indicated both for the first phase of measurement (Fig. 6.7a - see the description in § 6.1.1) and for the second phase of measurement (Fig. 6.7b). The second phase was introduced to obtain the values for evaluating the flow, whereby the readings from the first phase are also utilized. The output values from both measurements are the results obtained after conversion of the measured time intervals of propagation of ultrasonic pulses between the transducers. These numbers may be correlated with the corresponding distances. The phase of a measurement in which a reading was obtained may be identified by the second digit used for correlation.

In the first phase of a measurement, the numbers are obtained from which the coordinates are calculated in the way described above. In addition, with reference to (4.1), the velocity of sound propagation, c_{01}, can also be evaluated, in the direction from transducer M_m towards transducer M_o, and furthermore the velocities c_x and c_y, in the directions towards transducers R_x and R_y.

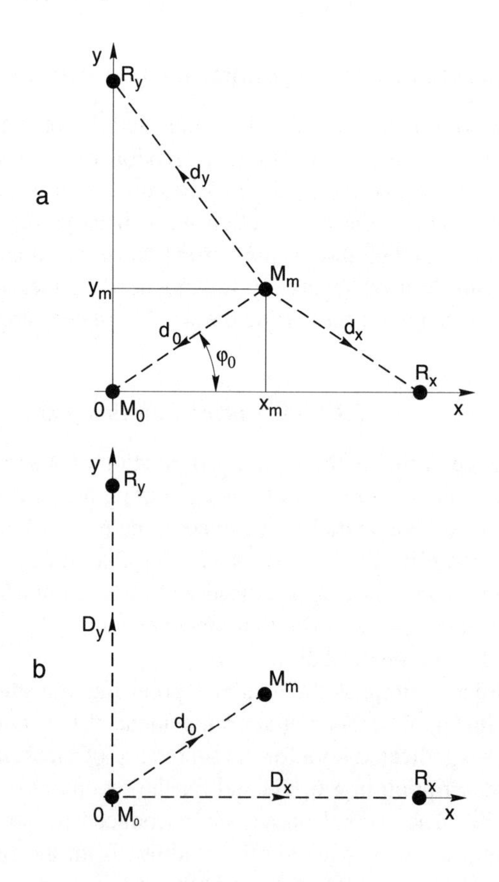

Fig. 6.7 ***Illustration of the arrangement of the point transducers, magnitudes of trajectories (moduli) and directions of propagation of ultrasonic pulses between them, in two phases of air flow measurement***

From the second phase of the measurement, when the transducer M_o is transmitting towards M_m, R_x and R_y, the velocities in the corresponding directions c_{o2}, c_x and c_y can be calculated. For a simple case, we can calculate the value of the flow velocity according to equation (6.5), if for the velocity, c_o, the average of c_{o1} and c_{o2} is substituted; similarly, for l, the value $d_{o1} = \sqrt{x_m^2 + y_m^2}$ is substituted. However, the last value is modified by an error from calculation of x_m and y_m. Moreover, the expression (6.5) is also only an approximation. A more precise formula can be obtained from reference [6.5], if we use the vector diagrams shown in Fig. 6.8. The flow vector, $\overline{v}_a$, is depicted there by a thick line as well as the sound velocities, measured in the corresponding directions, between the transducers. The other directions are depicted by thin lines. The value c_o is the velocity of sound in the stationary state (without flow), and it represents the radius of a circumscribed circle. c_o is appended to these vectors, to increase the facility of derivation of the expressions. Figure 6.8a corresponds to the first phase of measurement, and Fig. 6.8b corresponds to the second phase. Using the cosine rule, one can write

$$c = c_{o1}^2 + v^2 + 2c_{o1}v\cos(\varphi_o - \varphi_v) = c_{o2}^2 + v^2 - 2c_{o2}v\cos(\varphi_o - \varphi_v),$$

hence

$$c_{02} - c_{01} = 2v\cos(\varphi_o - \varphi_v) \qquad (6.8).$$

Furthermore, one may write

$$c_{02} = c_X^2 + v^2 - 2c_Xv\cos\varphi_v = c_Y^2 + v^2 - 2c_Yv\sin\varphi_v,$$

hence

$$c_X^2 - c_Y^2 = 2v(c_X\cos\varphi_v - c_Y\sin\varphi_v) \qquad (6.9).$$

By dividing expression (6.8) by (6.9), and rearranging, we obtain a formula for calculating the direction of flow

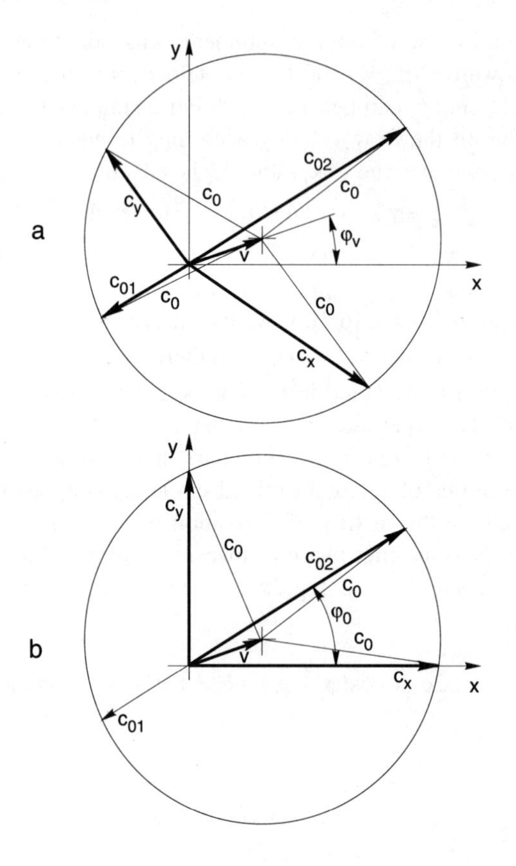

Fig. 6.8 ***Vector diagrams of the air flow velocities and the ultrasonic pulse propagation, in two phases of the air flow measurement***

$$\varphi_v = \arctan \frac{c_X(c_{02} - c_{01}) - (c_X^2 - c_Y^2)\cos\varphi_0}{c_Y(c_{02} - c_{01}) - (c_X^2 - c_Y^2)\sin\varphi_0} \qquad (6.10),$$

and, by applying (6.8), we obtain the modulus of vector

$$v = \frac{c_{02} - c_{01}}{2\cos(\varphi_0 - \varphi_v)} \qquad (6.11a),$$

or, by applying (6.9),

$$v = \frac{c_X^2 - c_Y^2}{2(c_X \cos\varphi_v - c_Y \sin\varphi_v)} \qquad (6.11b).$$

From (6.10), the angle φ_v can be obtained only in the interval (-90°, 90°). If the resultant velocity is negative, then the angle is in the third or the fourth quadrant. In this case, an angle of 180° must be added to the calculated value of the angle. If the denominator in (6.11a) is zero, then expression (6.11b) should be used. Assuming that c_o » v, the value obtained from the first measuring phase can be substituted for φ_o.

Similarly, the difference in velocities in the direction φ_o can be expressed as

$$c_{02} - c_{01} = c_{01}N_{01}/N_{02} - c_{01} \cong f\sqrt{x_m^2 + y_m^2}\,(N_{01} - N_{02})/N_{01}N_{02}.$$

After calculating the terms of the flow velocity vector, we can proceed with the correction of coordinates. The numbers N_{01}, N_x and N_y, measured in the first phase, are corrected to numbers N_{01k}, N_{xk} and N_{yk}, and the coordinate calculation is repeated again with these numbers, using the formulae given above (6.2). The corrected numbers are obtained from the relations

$$\frac{1}{N_{xk}} = \frac{1}{N_x} - \frac{v}{f\sqrt{(D_x - x_m)^2 - y_m^2}}\cos\left(\varphi_v - \arcsin\frac{y_m}{\sqrt{(D_x - x_m)^2 + y_m^2}}\right)$$

$$\frac{1}{N_{yk}} = \frac{1}{N_y} - \frac{v}{f\sqrt{(D_y - y_m)^2 + x_m^2}}\sin\left(\varphi_v - \arcsin\frac{x_m}{\sqrt{(D_y - y_m)^2 + x_m^2}}\right)$$

$$\frac{1}{N_{01k}} = \frac{1}{N_{01}} - \frac{v}{f\sqrt{x_m^2 + y_m^2}}\cos\left(\varphi_v - \arccos\frac{x_m}{\sqrt{x_m^2 + y_m^2}}\right) \qquad (6.12).$$

These correction relations are relatively complicated. For a very small velocity of flow compared to the velocity of sound, or if lower precision of calculation is sufficient, the relations for the calculation of the flow vector can be simplified, and an approximate correction of co-ordinates can be introduced. Then the simplified relations can be written, with reference to Fig. 6.9,

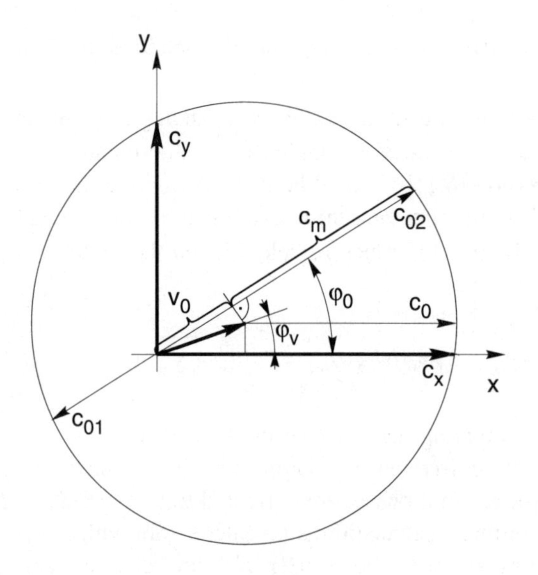

Fig. 6.9 ***Vector diagram illustrating the principle of simplification of the air flow vector calculation***

$$c_X \cong c_o + v\cos\varphi_v \cong c_m + v\cos\varphi_v$$
$$c_Y \cong c_o + v\sin\varphi_v \cong c_m + v\sin\varphi_v\,.$$

From these equations, the formula for the angle can be derived

$$\varphi_v = \arctan\frac{c_Y - c_m}{c_X - c_m} \qquad (6.13),$$

and another one for the velocity

$$v = \frac{c_X - c_m}{\cos\varphi_v} = \frac{c_Y - c_m}{\sin\varphi_v} \quad (6.14).$$

It is assumed that the velocity of sound, c_o, is approximately equal to the average sound velocity, c_m. This can be evaluated as

$$c_m = f\,\frac{\sqrt{x_m^2 + y_m^2}}{N_{01}}\ .$$

The value of the angle is within the interval (-90°, +90°). If the calculated velocity is negative, the angle must be increased by a value of 180°.

The correction of coordinates can be made in a simpler way than by correction of the numbers which were measured in the first phase using equation (6.12). The correction, and the subsequent calculation of coordinates, can be avoided by a method which is demonstrated graphically in Fig. 6.10. The procedure for direct correction of the coordinates calculated in the first phase is as follows: from the position vector, $\mathbf{A}_m(x_m,y_m)$, the error vector, $\mathbf{A}_c(x,y)$, is subtracted, hence a corrected

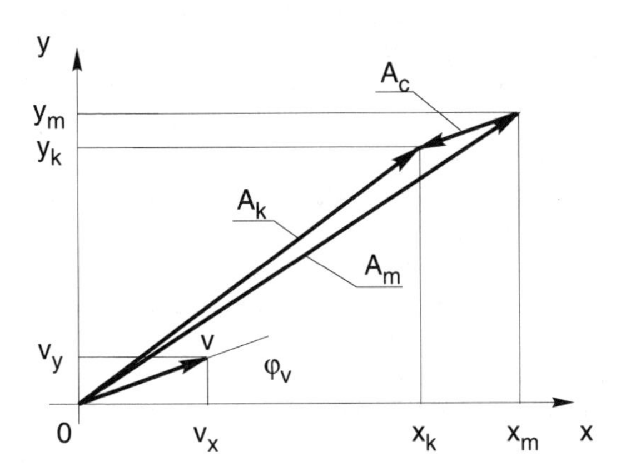

Fig. 6.10 ***Illustration of the principle of direct correction of the coordinates***

position vector $\mathbf{A}_k(x_k,y_k)$, is obtained. The error vector represents a shift in the measuring pulse caused by air flow during the measurement.

The terms for the position deviation vector can be expressed as

$$\Delta x = v_x t_o = v t_o \cos\varphi_v$$
$$\Delta y = v_y t_o = v t_o \sin\varphi_v \qquad (6.15)$$

where
$$t_o = \frac{t_{01} + t_{02}}{2} = \frac{N_{01} + N_{02}}{2f}$$

The corrected values of the coordinates are then given by the formulae

$$x_k = x_m - \Delta x$$
$$y_k = y_m - \Delta y \qquad (6.16).$$

6.3.2 Measuring flow in space

The spatial configuration of the (spherical) transducers is illustrated in Fig. 6.2c. Similarly to planar measurement (Fig. 6.7), and likewise in this (3D) case, the transducers can be considered to be point-like (Fig. 6.11). In the first phase of measurement, by digitizing the distances between the transducers, the numbers N_x, N_y, N_z and N_{o1} are obtained, corresponding to the duration of the pulse propagation from M_m over the corresponding distances to the transducers R_x, R_y, R_z and M_o. In the second phase, similar numbers N_x, N_y, N_z and N_{02} are obtained, corresponding to the duration of pulse propagation from the reference transducer, M_o, to the individual transducers on the coordinate axes.

The formulae for the individual terms of the air flow vector [6.6, 6.12] can be derived by the procedure described in section 6.3.1. For an approximate calculation, we obtain

$$\varphi = \arctan[(c_Y - c_m)/(c_X - c_m)]$$
$$\theta = \arctan[(c_Y - c_m)/(c_Z - c_m)\sin\varphi] = \arctan[(c_X - c_m)/(c_Z - c_m)\cos\varphi]$$
$$v = (c_X - c_m)/\sin\theta\cos\varphi = (c_Y - c_m)/\sin\theta\sin\varphi = (c_Z - c_m)\cos\theta \qquad (6.17)$$

The angles are marked in Fig. 6.11. In contrast to Fig. 6.7, the distances and the directions, for the first as well as for the second phase of measurement, are marked there. For c_m, an approximate value can be substituted,

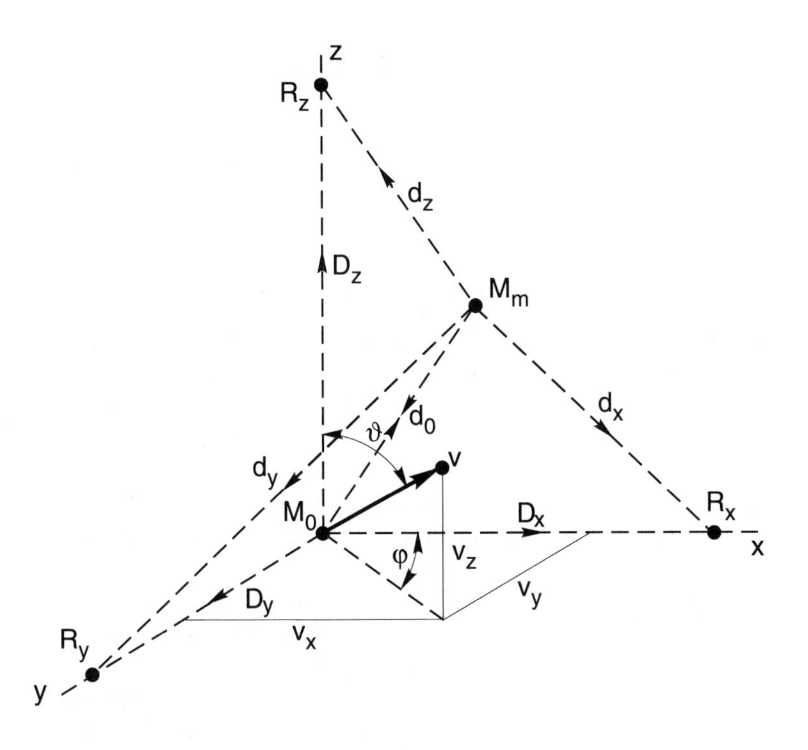

Fig. 6.11 ***Configuration of point transducers, distances and directions of propagation of ultrasonic pulses between them, for measurement of spatial (3D) flow***

$$c_m = f\,\frac{\sqrt{x_m^2 + y_m^2 + z_m^2}}{N_{01}}$$

or alternatively the average velocity may be used, $c_o = (c_{01} + c_{02})/2$, calculated from two measurements in opposite directions, between M_m and

M_o. Alternative relations are presented for the cases when the denominator is zero.

If an appproximate correction is considered satisfactory, the corrected coordinates can be calculated from

$$x_k = x_m - \Delta x = x_m - vN_{01}\sin\theta\cos\varphi/f$$
$$y_k = y_m - \Delta y = y_m - vN_{01}\sin\theta\sin\varphi/f$$
$$z_k = z_m - \Delta z = z_m - vN_{01}\cos\theta/f \qquad (6.18).$$

More precise formulae for calculation of the flow vector and correction of the coordinates are presented in [6.6].

6.4 Measurement of wind velocity

In the earlier sections of this chapter, an application of this instrument for coordinate measurement was described. Measurement of air flow was introduced for the purpose of an additional correction to the coordinate measurement, on the basis of values obtained in the second, complementary phase of the measurement. The precision of calculation of the separate air flow vector terms was influenced by the fact that the coordinates were not known beforehand. Their calculation was modified by an error to which the air flow also contributed.

Another situation occurs when the coordinate measurement is omitted, and the (originally moving) transducer, M_m, is located in a position with known, suitably chosen coordinates. Such an arrangement enables measurement of the velocity of air flow, i.e. the wind, for meteorological purposes and, if necessary, for other purposes as well. The dimensions of the equipment can be reduced because the configuration of the transducers and their arrangement are no longer constrained by the detected position. Moreover, the characteristics of the transducer need not be omnidirectional because the position of the transducer, M_m, does not alter. However, the requirement that the transducers M_m and M_o are reciprocal remains, because in this method, measuring the propagation time of an acoustic pulse in both directions between the transducers is required. As an extra, the air temperature can be also evaluated from the measured values.

6.4.1 Measurement of the velocity of wind in a plane

In section 6.2.1, a relation (6.5) was derived for calculation of the velocity of air flow, in a direction joining two transducers. If the direction of the flow is different, then only the projection of the velocity term in the relevant direction is measured. This time, the actual flow direction must be known, or at least assumed. In many cases, it is also necessary to measure the direction of the flow.

Let us suppose that the air flow has a planar, stationary character, i.e. its direction and size are equal in a measured plane. For this example measurement the transducers can be arranged according to Fig. 6.1 in such a way that $D_x = D_y = D$ and $x_m = y_m = D/2$. Then the relations (6.10) and (6.11) can be rearranged into the forms

$$\varphi_v = \arctan \frac{\frac{1}{N_X}\left(\frac{1}{N_{02}} - \frac{1}{N_{01}}\right) - \left(\frac{1}{N_X^2} - \frac{1}{N_Y^2}\right)}{\frac{1}{N_Y}\left(\frac{1}{N_{02}} - \frac{1}{N_{01}}\right) + \left(\frac{1}{N_X^2} - \frac{1}{N_Y^2}\right)} \qquad (6.19)$$

$$v = \frac{f D_X (N_{01} - N_{02})}{\sqrt{8}\cos(45 - \varphi_v) N_{01} N_{02}} = \frac{f D_x (N_Y^2 - N_X^2)}{2 N_X N_Y (N_Y \cos\varphi_v - N_X \sin\varphi_v)}$$

(6.20).

If the denominator of the first fraction in (6.20) is zero, the second expression should be used. If the resultant velocity is negative, the calculated angle should be increased by 180°.

The formulae for calculation of the direction and the velocity of wind can also be derived from the simplified relations (6.13) and (6.14). Given the configuration of the transducers described above, we obtain rearranged formulae of the form

$$\varphi_v = \arctan \frac{N_X(N_{01} + N_{02}) - \sqrt{2} N_X N_Y}{N_Y(N_{01} + N_{02}) - \sqrt{2} N_X N_Y} \qquad (6.21)$$

$$v = \frac{fD_X\left(N_{01} + N_{02} - \sqrt{2}N_X\right)}{N_X\left(N_{01} + N_{02}\right)\cos\varphi_v} = \frac{fD_X\left(N_{01} + N_{02} - \sqrt{2}N_Y\right)}{N_Y\left(N_{01} + N_{02}\right)\sin\varphi_v}$$

(6.22).

The same remarks hold true as for formulae (6.19) and (6.20). With regard to simplifying the measuring conditions by stationary positioning of the transducer M_m, the difference between the exact formulae and the approximate formulae is reduced. In spite of that, however, the maximum error in the approximate calculation compared to the more precise one is greater by about an order of magnitude. For example, for a wind velocity of $v = 10$ m/s (36 km/h), the error caused by simplification of the formulae is 4.1 %. For greater accuracy, or for higher velocities, though, the more precise formulae are recommendable. It is best to use a processor capable of calculating to a rather large number of significant places due to the occurrence of the differences of very small numbers or even their squares, especially in the calculation of angles. Thus the error in the angle calculation is transferred into the velocity calculation. In the conversion of distances to numbers, a high frequency, f (of the order of tens of MHz), for the clock pulses is often used in order to lower the discretization error.

6.4.2 Measurement of the velocity of wind in 3D

Again, the starting position is from the relations, previously derived, for the direction and the velocity of air flow in section 6.3.2, and for the same conditions as in the case of planar coordinate measurement, i.e. $D_x = D_y = D_z = D$ and $x_m = y_m = z_m = D/2$.

For an exact calculation, the formulae become rather complicated. These are presented in references [6.6] and [6.15]. For simplified calculation, they are rearranged from (6.17), as

$$\varphi = \arctan\frac{N_X\left(N_{01} + N_{02}\right) - 2N_XN_Y}{N_Y\left(N_{01} + N_{02}\right) - 2N_XN_Y}$$

$$\theta = \arctan \frac{N_Z(N_{01} + N_{02}) - 2N_Z N_Y}{[N_Y(N_{01} + N_{02}) - 2N_Y N_Z]\sin\varphi}$$

$$v = \frac{fD_X(N_{01} + N_{02} - 2N_Z)}{N_Z(N_{01} + N_{02})\cos\vartheta} \qquad (6.23).$$

Due to the non-linearity of the relations, the distribution of errors is non-uniform. The highest values of error occur at extreme positions and angles. For a wind velocity of 1 m/s (3.6 km/h), the maximum error caused by simplification is 0.4 %, while for a velocity of 10 m/s (36 km/h), it is already 4 %. When calculating using the more precise formulae, for a wind velocity of 10 m/s, the maximum error does not exceed 1.2 % (and elsewhere even less by an order of magnitude).

6.5 References

6.1 *Kéry M.:* Spôsob číslicového merania priestorových zložiek vektora okamžitej rýchlosti vetra a zariadenie na jeho prevádzanie. Pat. ČSSR No. 204 166, 1977 (in Slovak)[1]

6.2 *Kočiš Š., Oravec J.:* Elektroakustický číslicový rovinný snímač súradníc. Pat ČSSR No. 222 855, 1980 (in Slovak)[2]

6.3 *Kočiš Š.:* Elektroakustický číslicový priestorový snímač súradníc. Pat. ČSSR No. 214 290, 1980 (in Slovak)[3]

6.4 *Kočiš Š.:* Electroacoustical Plane and Spatial Coordinate Measurement Methods. Elektrotechnický časopis, 32, 1981, No. 5, 585-597

6.5 *Kočiš Š.:* Korekcia vplyvu prúdenia vzduchu na chybu merania súradníc elektroakustickou metódou. Elektrotechnický časopis, 33, 1982, No. 10, 786-796 (in Slovak)[4]

6.6 *Kočiš Š.:* Electroacoustical Robot Grip Position Measurement. Computers and Artificial Intelligence, 2, 1983, No. 5, 479-488

---

[1] A method for digital measurement of the 3D component terms of an instantaneous velocity wind vector and a device for performing this

[2] An electroacoustic digital sensor for planar coordinates

[3] An electroacoustic digital sensor for spatial coordinates

[4] A correction for the effect of air flow on the error in coordinate measurement by an electroacoustic method

6.7 *Kočiš Š.:* Ultrasonic Pulse Method and its Application for Numerical Plane and Space Measurement. Elektrotechnický časopis, 35, 1984, No. 11, 840-851

6.8 *Rudashevskii G.E., Zenin V.Ya.:* O vliyanyi dvizhenyia vozdushnoi sredy na tochnost izmerenyia rasstoyanii v mashinovedenii s pomoshchyu ultrazvuka. Mashinovedenyie, 1985, No. 2, 59-62 (in Russian)[5]

6.9 *Kočiš Š., Oravec J., Toman M.:* Ultrasonic Numerical Spatial Measurement of the Robot Arm Movement. Proc. Xth World Congress IMEKO '85, Preprint, Vol. 5, Dom Techniky, Prague, 1985, 236-242

6.10 *Chaude P.K., Sharma P.S.:* Ultrasonic Flow Velocity Sensor Based on Picosecond Timing System, IEEE Trans. Ind. Elec., 33, 1986, No. 2, 162-165

6.11 *Kočiš Š., Bystriansky P.:* Zariadenie na ultrazvukové číslicové meranie smeru a rýchlosti prúdenia plynného alebo kvapalného média v rovine. Pat. ČSSR, 1987, No. 260 192 (in Slovak)[6]

6.12 *Kočiš Š.,Bystriansky P.:* Zariadenie na ultrazvukové číslicové meranie smeru a rýchlosti prúdenia plynného alebo kvapalného média v priestore. Pat. ČSSR, 1987, No. 260 193 (in Slovak)[7]

6.13 *Kočiš Š., Bystriansky P.:* Elektroakustické číslicové meranie rýchlosti a smeru vetra. In: Proc. Int. Conf. EMISCON '87, Poster section, Dom Techniky, Košice, 1987, 76-78 (in Slovak)[8]

6.14 *Toman M., Bystriansky P.:* Generovanie a snímanie ultrazvukových impulzných vĺn pri meraní vzdialeností, súradníc a parametrov prostredia. Elektrotechnický časopis, 40, 1989, No. 2, 127-140 (in Slovak)[9]

6.15 *Bystriansky P.:* Meranie prúdenia vzduchu a korekcia jeho vplyvu pri elektroakustickom snímaní súradníc. Thesis, Electrotechnical faculty, Slovak Technical University, 1992, 115 pp (in Slovak)[10]

6.16 *Commercial Press:* Weather Measure Corp., Sacramento, U.S.A., Scientific Instruments & Systems Catalog 1087. W 115 Sonic Anemometer-Thermometer

[5]The influence of movement of air on the precision of distance measurement in leading machines by ultrasound

[6]A device for ultrasonic digital measurement of direction and velocity of flow of a gaseous or a liquid medium in a plane

[7]A device for ultrasonic digital measurement of direction and velocity of flow of a gaseous or a liquid medium in 3 dimensions

[8]Electroacoustic digital measurement of velocity and direction of wind

[9]Generation and detection of the ultrasonic wave pulses for measuring distance, coordinates, and medium parameters

[10]Measurement of air flow and correction of its influence on the electroacoustic sensing of coordinates

6.17 *Commercial Press:* Kaijo Denki Co. Ltd., Japan: Digitized Ultrasonic Anemometer Thermometer, Model DAT-300. Prospectus, 26 pp.

7 Ultrasonic instruments: case studies

7.1 Two-dimensional sensors for planar coordinates

The most frequently used application of ultrasound as a measurement technique is for measurement of distance and position. A considerable advantage of ultrasonic measurement of position is that it is possible to measure more coordinates at a time using a single source of ultrasonic waves. This advantage is often used in instruments for measuring the coordinates of graphical patterns. A name for these two-dimensional graphical digitizers has been developed. They are designed mostly as graphical input peripherals for computers. Thus they are matched for interfacing to a computer, which can control them and which receives data from them corresponding to the measured coordinates. This indicates their wide application, for instance in CAD/CAM systems, in cartography, architecture, building engineering, medical diagnostics and other applications.

The sources for generating ultrasonic waves have been described in Chapter 3. For measuring planar patterns, spark sources are most often used which generate spherical ultrasonic waves and also ring-shaped piezoelectric transducers which generate a cylindrical ultrasonic wave.

In the case of acoustic-electric transducers, i.e. microphones, the choice of physical principles for conversion is slightly wider. Recent developments aim towards the improvement of reciprocal transducers. Of these, the capacitor and the piezoelectric transducers are most often used. Their design is either extended (e.g. linear transducers), or point or quasi-point like. Recent developments aim towards the second type, however the design is very demanding, especially in the case of the reciprocal type.

A survey of some methods utilized for coordinate measurement is given in Chapter 4. A more detailed description of one of these, specified mainly for measuring planar and 3D coordinates in very demanding environments, is presented in Chapter 6. In this section, we will describe two typical models of planar coordinate sensors for graphical patterns produced by the Science Accessories Corporation (SAC). This company has dealt for a long time with the development and production of graphic digitizers [7.1 - 7.5]. Amongst their range there are also models for 3D measurements.

Sonic digitizer, model GP-9

The basis of the digitizer is a control box, elongated linearly in such a way that on both its ends, two miniature (point-like) microphones, A and B, can be fastened (Fig. 7.1). Data capture is accomplished by the method indicated in Fig. 4.16, and described in the corresponding text. For compensation of atmospheric effects, a reference cylindrical piezoelectric transducer, C, is appended to the structure, placed at a distance of about 10 cm in front of transducer B.

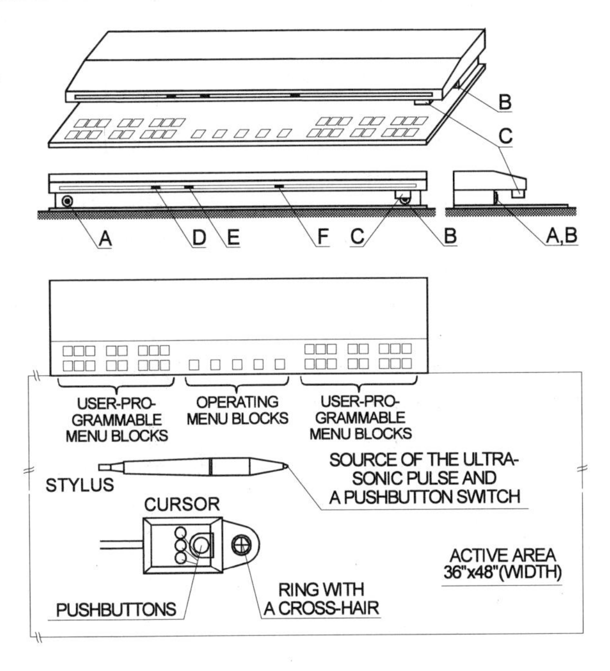

Fig. 7.1 ***An angle view, a front view and a side view of an ultrasonic digitizer SAC, model GP-9***

In the front of the cursor a ring-form piezoelectric ultrasound emitter is built-in with a cross-hair inside. It can move over the working area of dimensions 36″ × 48″ (90 × 120 cm), and in the case of the special model GP-9 XL, the dimensions are 40″ × 60″ (100 × 150 cm). On the front panel, 3 LED indicators are placed.

The digitizer is operated directly by a multi-purpose menu from the computer. The menu is recalled (initialized) by activation of the cursor, which is positioned above the corresponding rectangle on the digitizer pad. Then signals are emitted to the control unit, which behave like invisible switches. The individual rectangles, according to their numbering, generate letters in ASCII code, which are interpreted by the software. The calculation of coordinates can be performed in several coordinate systems. Most often a rectangular cartesian system is used in which any point is determined by a pair of coordinates, x and y. The origin of the coordinate system can be shifted to any point of the active area. It is usually positioned in the left lower corner. Then the whole active area represents the first quadrant.

The menu items allow selection of one of three ways of sensing: points, lines and flows. A special batch mode enables digitizing of each point eight times in one sequence. The accuracy is increased by averaging these eight measurements.

Before each coordinate measurement, a calibration is performed. During the calibration, the actual velocity of propagation of the ultrasonic pulse is measured. Following its evaluation, a correction to the measured coordinates is carried out. In this way the effect of atmospheric fluctuations is suppressed. For this purpose, the reference transducer, C, is used, with a known position relative to the microphones.

The corrected coordinates are calculated by a dedicated microcomputer at a rate up to 100 pps (points per second). The data are then transferred to a host computer for further processing. The corrected coordinates are also shown on the CRT monitor.

A check on the validity of the data is performed by software which determines how to treat the wrong data. If the data check is on, each point is digitized twice (in an interval of 15 ms). If there is, for instance, unusually large interference, then the two digitized values of a single point may not be equal. These are compared and their further treatment is decided by the application software.

The model GP-9 uses a 5-point weighted floating average for smoothing the data in the trajectory and flow modes. This increases the stability of a measurement.

The conditions of use of the ultrasonic digitizer are limited by propagation of the ultrasonic pulses in the measurement volume. For proper operation, the requirement that there are no obstacles between the ultrasonic transmitter and the receivers, which would prevent the propagation of an ultrasonic wave in the shortest direction between them, should be unconditionally satisfied. At a higher frequency of pulse repetition, interference may arise between the transmitted and the reflected ultrasonic waves from nearby obstacles. On digitization above a surface that is sensitive to the digitizing frequency, vibrations can arise which may cause a failure of the measurement. This effect can be suppressed by placing the surface frame on a pad made from a different material.

Automatic calibration currently compensates for the influence of fluctuations of the atmospheric parameters; however this is only if they are distributed uniformly. An unequal temperature distribution (a temperature gradient) leads to an auxiliary temperature error. This occurs, for instance, when a temperature source is placed close to the instrument. In addition, air circulation can appear, which is not fully compensated and contributes to the resultant error. The air flow can also arise for other reasons, for example opened doors, windows, effects of air conditioning, etc. Environmental sources of noise can also affect the reliability of a measurement, especially if the noise contains frequencies above 20 kHz.

If proper measuring conditions are maintained, digitization can be performed with a resolution of 0.005″ (0.125 mm), and with a repetition rate up to 38400 Baud. The working temperature of the instrument is within the range 32-120 °F (0-49 °C).

7.2 Robot grip position tester

A characteristic attribute of a robot is its ability to move from a given to a new position within its operating space along a prescribed trajectory.

According to reference [7.6], as far as positioning is concerned, the reproducibility of most robots is of the order of mm, and, for more precise machines, even of the order of 0.1 mm or better. However, if the absolute position should be determined, errors of even several cm can often arise, especially if the robot operates at its maximum speed. This is caused by the superposition of errors from many different factors, ranging from poor knowledge of the kinematic parameters of the robot, up to the elastic deformation of the robot arms, and the elongation due to temperature.

A demand for minimal tolerances in the production of robot components leads to the use of special materials, and increasing stiffness of the arms and joints, which is not always economical. Often the utilization of external sensory systems for position sensing can be more effective. The internal sensors of the individual robot joints provide only partial information and their superposition may not be sufficiently precise for the determination of an instant position of the robot's end point. In that case a change-over is necessary from incremental sensors to absolute position sensors. Moreover, the application of an external sensory system ([7.7], [7.8]) significantly improves the calibration of the robot.

Implementation of the sensor subsystem (SS) into a robotic system inevitably requires the integrated design of a robot. Only then can the maximum speed of sensing be achieved as well as the processing and transmission of the measured data. Then the speed is limited only by the physical measuring principle used and thus by the limiting velocity of propagation of an ultrasonic wave in air.

This limit can be overcome by using special techniques, as for example by emission of a sound wave earlier than the front face of the previous sound wave was detected by the sensor. In this way, the repetition frequency of measurement can be further increased, by about 3- to 5-fold.

The main objective of an ultrasonic sensory system is to elaborate the interaction of a robot with its surroundings. The sensory system alone does not provide a sufficiently precise determination of position in real time, so that it cannot be the sole source of positional information about the robot's working space. However, specifying its position within the working space (or even of a group of robots) provides important information, which in combination with other sensors (e.g. proximity sensors) can mean a considerable contribution to the robot position control (see references [7.7], [7.8]). Considering that the instants of position sampling are well defined, the sensory system also enables determination of the instantaneous velocity, as well as acceleration of the robot's end point.

When combining several sensors with different physical principles of operation, it is possible to suppress the drawbacks of the individual pick-ups that are integrated into the sensory subsystem of a robot. Recently many research teams have worked on this problem ([7.9], [7.10], [7.11]). For this task the term *sensor fusion* or *data fusion* is often used.

When using an ultrasonic sensory system, it is interesting to combine it with another system which enables the capture of more precise information in the cases when the use of ultrasound exhibits drawbacks. This can be:

- in critical parts of the working space of this sensory system (increasing some errors in the method)
- in close proximity to the manipulated objects

The information from an ultrasonic sensory system can be conveniently complemented by optical information. If the problem of sensor integration is suitably solved, the outcome represents a sensory subsystem with increased reliability and precision, compared to the separate sensors alone. This is due to the fact that the information-carrying signals of individual sensors are mutually correlated, while the noise in these sensors is uncorrelated.

Sensor fusion (a careful and thoroughly designed integration of sensors), possibly in addition with satisfactory redundancy of information, provides a certain robustness to the output data obtained. This effect is well known in technical practice, in control of aircraft, as well as in the organisation of the human sensory system; however, in robotics, up until now it has been used only in top quality systems.

If the information content of the measured spatial coordinates (at the precision and speed of an ultrasonic sensor system actually achieved) can contribute to efficient robot control, it is necessary to account for this fact already in the conceptual design phase of a particular robot and its control system. In such a case, it is necessary:

a) to integrate the cursor of the sensory system into the robot gripper mechanically
b) to place the electronic circuitry, power supplies and the wiring cables of the sensory system suitably
c) to solve the problem of proper connection of the robot control system with the sensory system by technical means, i.e. by alloting at least one communication channel for this purpose
d) to choose a suitable configuration of measuring channels for the sensory system with respect to the operational space of the robot
e) to solve the question of pre-processing the measured values obtained from the sensory system, in order that there is the possibility in the control system of evaluating the robot position data in real time

f) to assign part of the capacity of the processor (or some of the processors) of the control system to the sensory system, and to extend the application programming language to obtain and utilize the absolute coordinates of selected important points.

The last two points relate to the software of the sensory system, or the supervisory control system of the robot.

The information on the position of the chosen extra (most important) point represents an input for an inverse kinematic problem of robot control. This information is significant and usable directly in an analytical solution for robot control. Off-line planning of a trajectory, for instance in automatic analytical programming, is important in creating robototechnical complexes, in linking the operation of individual robots into a firm sequence, etc. In this field, sensors for determination of absolute position in the working space play an irreplaceable role.

In some categories of robots (e.g. mobile and spraying robots), the precision of an ultrasonic sensor system is sufficient for determination of their position. The aim of robot testing is the determination of the coordinates of the most important (extra) point (or points) in space, for prescribed testing positions and movements.

The positioning accuracy of a robot means its ability to move (i.e. change position) to the appointed place within its working space, within an allowable error. The precision is understood as the error in repeated measurements in moving to a given goal from a given initial position. Modern robots ensure a good repeatibility, however their accuracy is considerably worse. In on-line robot programming (i.e. learning), repeatibility is an adequate specification. However, with the appearance of computer-assisted-manufacture (CAM), industrial robots must be programmed off-line. The problem-oriented languages are used, which are able to transform information from a CAD/CAM database into instructions for the movement of industrial robots. This way of programming, based on an analytical prescription of position, already requires high precision of the robots.

For the process of calibration of robots in industry, or periodically, while they are in operation, it is necessary to fulfil the following steps using an ultrasonic testing stand:

1) modelling a relationship between robot parameters and the position of the working tool

2) testing measurements
3) identification of error sources, and
4) correction of the model according to point (1).

The testing equipment for robots is often based on equipment used in the machine industry for automated testing of precision during production. However, their direct application is usually impossible because of the specific character of robots, above all in relation to their dynamic character.

The advantages of an ultrasonic sensory system for measuring coordinates are mostly that

- the sound propagation velocity enables a direct, absolute measurement of distances (by the PTM, Propagation Time Method), as well as a relative measurement by phase evaluation (PDM, Phase Difference Method) [7.12]
- all three cartesian coordinates corresponding to the instant of sampling can be determined by processing the numerical data
- it has relatively good dynamic characteristics as the mass of the miniature source of an ultrasonic wave pulse is negligible relative to the mass of the robot effector.

In sensory systems, after the suppression of scalar effects (which are already involved in the basic measuring method), it is also possible to solve satisfactorily the question of the influence of the air flow in the working space of the testing equipment.

The limitations on practical use are determined by these parameters:

- the precision of repeated position measurements for all three coordinate terms is better than 0.1 mm
- the number of measurements within the working space of a cube of side 1 m is at least 100 measurements per second.

Without a more complicated process of calibration, the testing equipment (Fig. 7.2) can be used for determination of the precision of positioning for repeated measurements, for determination of the precision of positioning from various directions, as well as for the error in tracing the trajectory. For the last parameter, the measurement utilizing the ultrasonic testing stand is the easiest one, and probably also the most advantageous, as the additional mechanical load on the robot arm due to the fastening of the sound source is negligible, compared with other testing methods.

For calibration of the whole system, the methods of direct setting using coordinate measuring machines come into consideration, or indirect calibration methods.

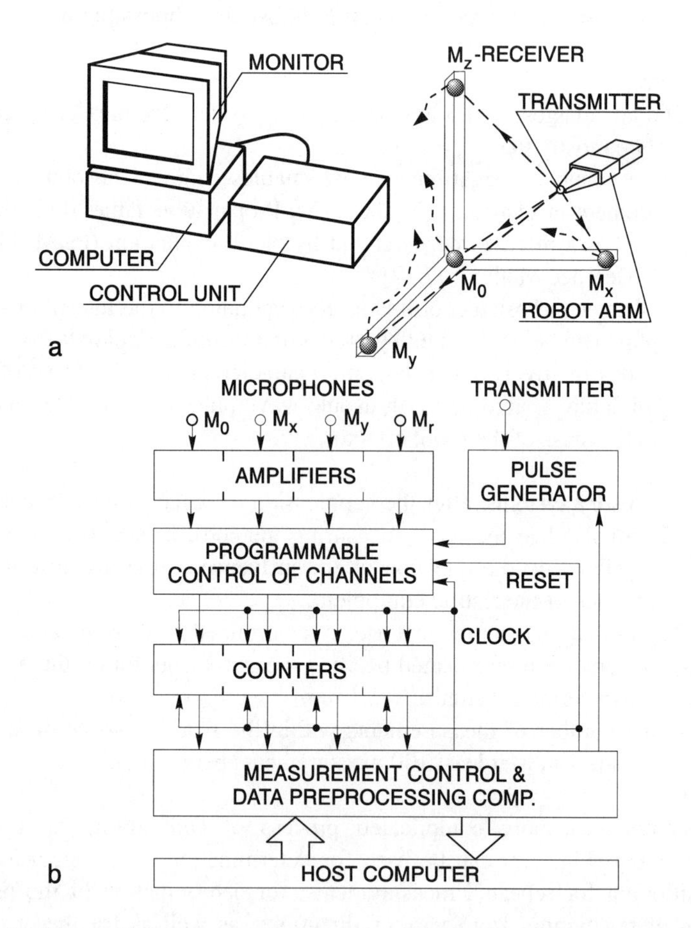

Fig. 7.2 ***Schematic diagram of part of a test area for testing the trajectory of robot arm motion***

Such a calibration can be accomplished by means of an analysis of the measured values, obtained from precise movements of a calibration robot, by a coordinate measuring machine, or by other precision equipment. The procedure leads to a calculation of precise, actual positions of the microphones. The positions, corrected in this way, should then be introduced into the equations for calculation of the coordinates of the measured points in space. Here the exploitation of artificial neural networks also finds application [7.10].

7.3 Instruments for measuring air flow

Instruments for measuring air flow are generally called anemometers. Several principles and methods are used for measurement, in rotational anemometers, thermoanemometers, ultrasonic gauges, laser methods, sonic radars, etc.

Ultrasonic anemometers are designed and produced commercially as precise and reliable digital measuring instruments. They are able to measure 1D- , 2D-, as well as 3D- vectors of the flow velocity of air, or other gaseous or liquid media.

The problem of measuring air flow has been described in section 6.2. The principle of air flow measurement in a 1D variant represents the basis of flow-meters (flow-gauges). By extension of the set-up of ultrasonic transmitters and receivers, and their suitable arrangement, instruments are developed for measurement of two or three components of the velocity vector - sonic anemometers. Typical representatives of this class are the products of the WMC company (Weather Measure Corporation) [7.15]. They are interesting in their ability to also measure the air temperature, besides the flow. If, in particular, information about the medium flow velocity is obtained from the difference of times t_1 and t_2 through relation (6.4), then from their addition, $(t_1 + t_2)$, information can be obtained about the medium temperature, ϑ (in Kelvin)

$$t = t_1 + t_2 = \frac{2\ell c_0}{c_0^2 - v^2} \cong \frac{2\ell}{c_0} \tag{7.1}$$

$$c_o = 20.067\sqrt{\vartheta} \tag{7.2}$$

$$\vartheta = \left(\frac{2\ell}{20.067t}\right)^2 \tag{7.3}$$

where ℓ - distance between transducers.

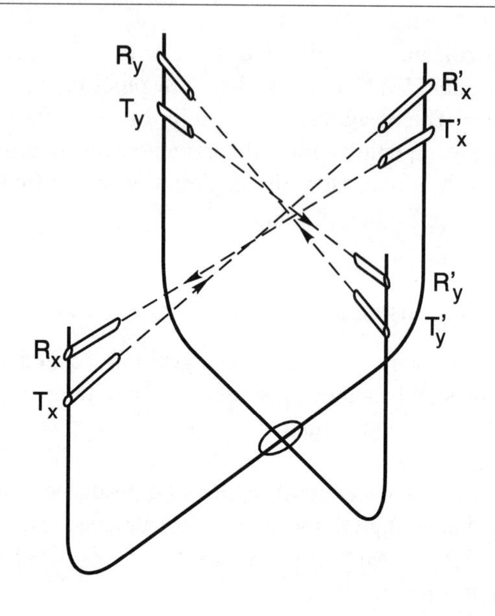

Fig. 7.3 ***An ultrasonic anemometer in a 2D setup***

In Fig. 7.3, a sketch of an arrangement of transmitters and receivers is shown for the 2D type instrument W 115, with separate ultrasonic transmitters and receivers for a planar wave pulse, on the basis of piezoceramics (see the measurement principle in section 6.2.1).

A good illustration of this type of anemometer/thermometer for meteorological purposes in 3D is the instrument DAT-300, - a product of the company Kaijo Denki Co. [7.16]. In the sensor part, three pairs of reciprocal ultrasonic transducers are used. An outline of the configuration of the sensing probe is shown in Fig. 7.4. The instrument achieves remarkable specifications: measurement of air flow velocity (the components v_x and v_y) within the range 0-30 ms^{-1}, measurement of the v_z component from 0-10 ms^{-1} with a resolution of 0.005 ms^{-1} and an accuracy of ± 1 %. The range of air temperature measurement is from −10 °C up to +40 °C, with a resolution of 0.025 °C, and an accuracy of ± 1 %. The individual measurements are carried out with a repetition frequency 20 Hz. The axes of individual pairs of ultrasonic transducers are skew in order that interference does not occur during the simultaneous excitation of all the transmitters.

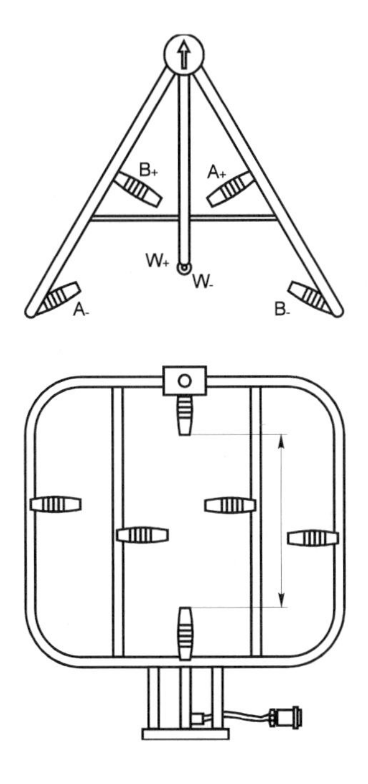

Fig. 7.4 ***A commercial 3D anemometer DAT-300***

A similar measuring instrument is described in reference [7.17]. This instrument serves for measuring the 3D vector of air flow velocity to military specifications.

In the next section, a description of an experimental anemometer is presented, which is able to measure the 2D velocity vector of air flow, the coordinates of a point in a plane, as well as the air temperature. It is a laboratory instrument, developed at the authors' place of work. The instrument was developed as an automatic measuring system based on a microcomputer. The sensory system uses reciprocal, cylinder-shaped, capacitor ultrasonic transducers and radial emission of ultrasonic waves (Fig. 7.5). It represents an enhancement of the principle illustrated in Fig. 6.5c. The calculation of direction and velocity of the flow is performed using formulae (6.10) and (6.11). The block diagram of the instrument is shown in Fig. 7.5c.

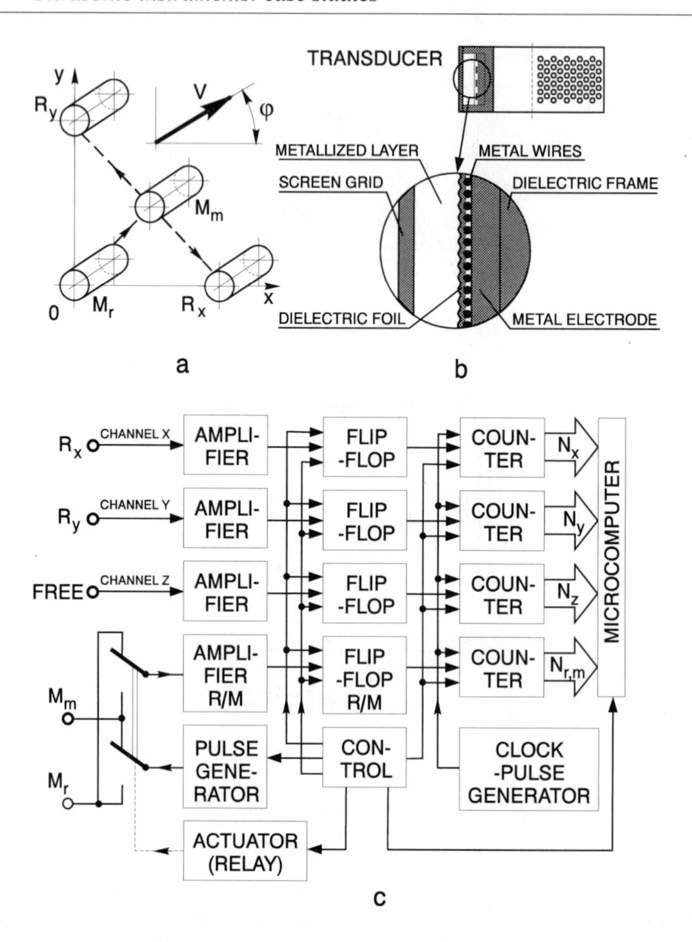

Fig. 7.5 ***Configuration of the transducers (a), an illustration of their arrangement and a cross-section of the active part (b), and a block diagram of the instrument (c), for ultrasonic measurement of coordinates, and of air flow***

The operation of the instrument was tested by taking measurements in a wind tunnel, in the velocity range 0.5-5 ms^{-1}, and in directions from 0 to 360°. The experimental results indicate that the instrument is suitable for measurement of low velocities, with little turbulence, and in surroundings without intense acoustic interference. An advantage of the instrument is its ability to measure coordinates, the air flow and air temperature at the same time. In order to perform such a complex task, the position coordinates of the transducer M_m, which is movable this time, are measured. The instrument can also be extended to 3D measurement. However, for such a task, it is necessary to use spherical or quasi-point transducers.

From the point of view of measuring precision, the transducers, especially the reciprocal ones (M_m, M_r) are important, as well as the associated analog electronic circuits. Cylindrical, capacitor-type transducers were chosen with a gold-plated styroflex foil of thickness 20 μm (Fig. 7.5b). A version with a thinner (6 μm) aluminized foil gives higher sensitivity in the microphone mode, however it has a low efficiency in transmission mode. The diameters of the active parts of the transducers were 20 mm and 40 mm.

7.4 Ultrasonic measurement of thickness

In this section, some practical examples of thickness gauges which are on sale at present are introduced, as well as a description of the possibilities which are offered by the ultrasonic measurement of thickness in industry.

The precise measurement of thickness has made a substantial step forward by the introduction of the latest instrument from the Wells Krautkammer company [7.18]. The new instrument, CL 304, reaches a resolution of 2.5 μm. For the first time, it is possible to measure ceramics. The instrument comes in handy in a laboratory as well as a workshop.

Microprocessor control leads to easier use and higher measuring power compared to the previous model, CL 204. A further improvement of the emitting and amplifying circuits has led to a very good resolution close to the surface, without energy losses due to penetration into the measured material.

The setting and preparation for stand-by state is very quick and easy. The instrument is reset merely by pushing a button, and the calibration is performed by means of a material of known thickness, or with the aid of the known velocity of sound propagation in the material. With the instrument CL 304, new horizons have been reached. Since the measured range of sound velocity is extended to 20 000 ms^{-1} (a two-fold increase of the standard maximum value), very fast, (i.e. acoustically quick) materials can also be measured, such as ceramics, as well as acoustically slow plastics.

The parameters of a measurement, such as gating, amplification and polarity, which are applicable for particular materials, thicker parts, etc., can be stored in the memory for future use. Up to seven sets of values can be stored, which saves time in changing the working mode.

For evaluation of the dimensional tolerance, differential mode is used, and the upper and lower limits, in connection with LED indication of failure, create a pass/fail condition which can be operated by inexperienced staff.

The standard output values, including high frequency signals, can be connected to an oscilloscope to elucidate the measurements in critical cases. This method is very suitable for testing parts which have a complicated shape and a large number of dimensions.

An interface enables connection of auxiliary instruments for data storage, a printer or a computer. This interface can be supplied later, as an option. Therefore the instrument CL 304 can also be used in an extended mode, for statistical process control, and for various checking and testing applications.

Instruments from other manufacturers also have similar features [7.19].

Cygnus Instruments Ltd. [7.20] use a new principle for digital thickness measurement in the CYGNUS 1 thickness gauge, which ignores various surface layers (spray, polish, etc.) on the material, and measures only the actual thickness of metal. The instrument has a torch shape and it operates on an echo-impulse principle with a single transducer. Instead of evaluation of thickness from the time interval between emission of the signal and the return of the first echo, the time count starts at the first detected echo, and the time intervals are measured between the first and the second echo, reflected from the rear side of the material (Fig. 7.6). Of course, the electronic circuits are more complicated because the echoes from other, interfering, reflections must be resolved and suppressed. The instrument detects at least three echoes before a reading appears on the display.

This mode of operation enables measurement of thick, corroded steel surfaces, unpolished and rough surfaces or surfaces with layers. The process is fully automatic and is very simple to manipulate.

The operating range is from 1.5 to 99.9 mm, with an accuracy of 0.1 mm. The standard (default) calibration is for mild steel, however, the instrument can be adjusted for velocities between 1 and 6.5×10^3 ms^{-1}, so that aluminium, copper, glass and other materials can also be measured.

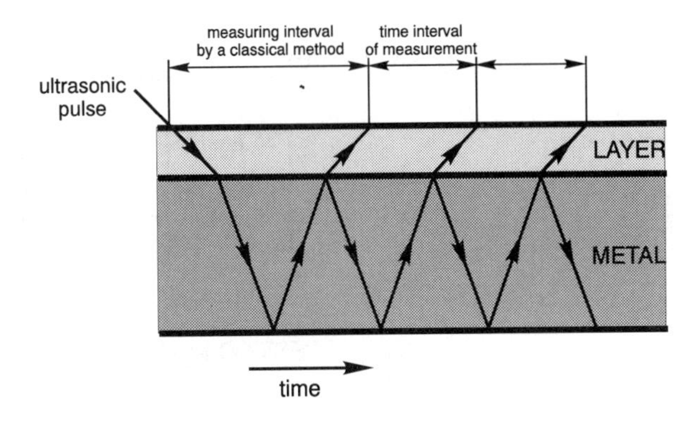

Fig. 7.6 ***The principle of operation of a thickness gauge, the CYGNUS 1***

The transducer is made from lead metaniobate and it works at a frequency of 2.25 MHz. As an option, a special probe can be ordered for measurements in difficult corners, thin pipes, etc.

Another example of thickness measurement using ultrasound is the measurement of silicon wafers during the process of lapping [7.21]. The thickness can be measured by detection of the duration of wave transitions through the sample. This can be expressed by the equation

$$l = 1/2\ vT \tag{7.4}$$

where l - material thickness
v - longitudinal velocity
T - duration of the wave transition through the sample.

This method is usually called a direct contact method, as the transducer is in direct contact with the measured sample (Fig. 7.7). It cannot be used simply for measuring thin plates such as silicon wafers, because the reflected pulse then overlaps with the emitted pulse, and moreover multiple reflection often occurs. To overcome this problem, an indirect contact method has been developed.

The method rests on a similar principle, only a delay element is used, which consists of a solid body, or a liquid column, placed between the transducer and the thin measured plate. This is why the method is also called a method of a solid delay line, or a method of a liquid delay line.

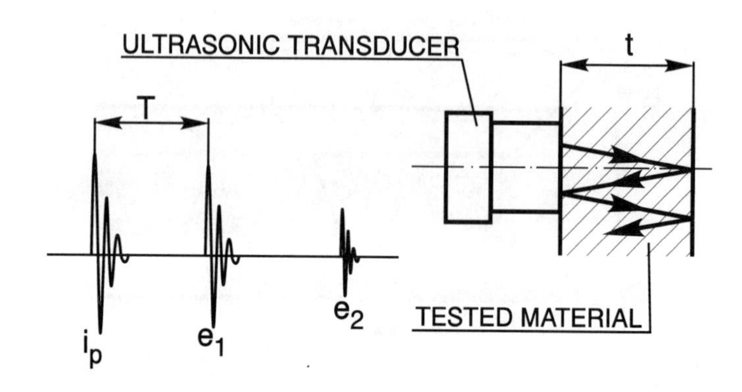

Fig. 7.7 ***The principle of thickness measurement for direct contact of an ultrasonic transducer with the material***

The outcome is that low-index reflections, b_i (i = 1, 2, 3,...n), are obtained, which appear between the first and the second surface reflections, as can be seen in Fig. 7.8. The time, T, can easily be measured in such a way that any (arbitrary) pair of successive reflections are selected, because every low-index reflection corresponds to a single pulse. Then the thickness can be calculated from equation (7.4).

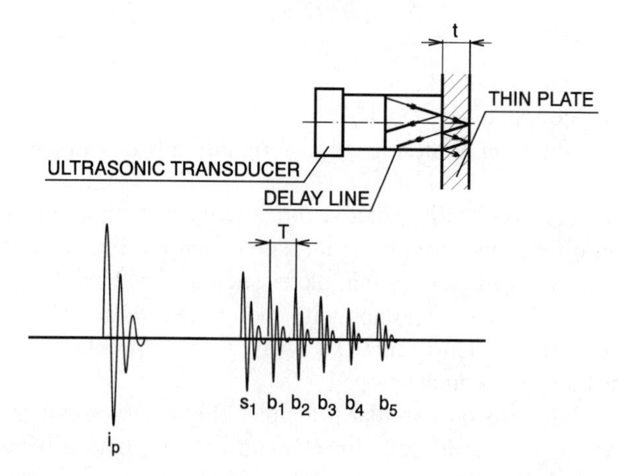

Fig. 7.8 ***The principle of thickness measurement using a delay line, for the case of a thin plate***

Since the delay line must be in contact with the wafer during the lapping procedure, a liquid delay line seems to be more suitable. From the practical point of view, a grinding sludge rather than pure water can obviously form the required column of liquid.

By introducing the method described into the process of automatic measurement of dimensions on the lapping machine, a scattering (variance) of 2.6 μm from the average thickness value was obtained. A 20 MHz ultrasonic probe was used for the measurement.

7.5 Distance and level measurement

Distance meters and level gauges are manufactured by various producers. In this section, some examples of such instruments are presented as well as examples of their application.

Recent ultrasonic sensors which recognize the presence of a reflecting surface at a given distance have become widespread. A variety of manufacturers [7.22 - 7.26] produce this type of detector. Their realization was made possible by the appearance of new materials for matching the acoustic impedance of transducers to the impedance of air.

Compared with instruments which need physical contact with the object whose surface is to be detected, position sensors have some outstanding advantages:

- they can be mounted beyond the expected range of positions of the object
- an early reaction can protect both the object and the sensor against a destructive collision
- the resistivity and precision of the sensor is not influenced by wear caused by physical contact (as occurs in tactile types).

These positional sensors can be realized on the basis of various physical principles. Inductive and the capacitive contactless limit switches create an ac electromagnetic or electrostatic field, and sense its deformation by the detected object. However, their use is limited to electrically conducting objects and their detection range is usually very small, 10 mm or less as a rule. Optical distance sensors, which are used for longer distance measurements, and which use reflections from the detected object, are rather sensitive to dust and dirt, and their dependence on the reflection coefficient is much larger than that of ultrasonic instruments. Ultrasonic sensors have a unique privilege in that they are sensitive to structures in the viewing direction, i.e. in the direction of propagation of the ultrasound. In addition, they recognize objects in a relatively well defined range, even at larger distances. Since transducers with working frequencies in the range 80 - 400 kHz have been developed, proximity sensors that are insensitive to dust and dirt have been produced. By using recently developed microelectronic components, the relative complexity of the necessary electronic circuitry no longer represents the price limiting factor. High resolution is achieved by selection of a high operational frequency. The high attenuation which occurs during the propagation of ultrasound in air limits the nominal detection range, however, at the same time, it eliminates efficiently acoustic interference.

With an ultrasonic position sensor, a pulse echo instrument is used which analyzes the delay of the reflected signal, so that the determination of the presence of objects, within a preselected detection range, is possible. As shown in Fig. 7.9, an ultrasonic transducer emits a pulse of about 30 waves at the working frequency (e.g. 200 kHz). This pulse, reflected by the object which is to be detected, returns as an echo back to the transducer, which in the meantime has been switched over to receiver mode. The echo pulses propagate with a time-dependent gain which compensates the drop of amplitude caused by attenuation with increasing distance. Using the same transducer for transmitting as well as receiving is helpful because it has the same resonance frequency and also the same directional characteristics in both modes. The minimum detectable distance, d_{min}, corresponds to the time interval in which the fading emitted signals interfere with the expected echo signals. The spatial detection range can be selected by limiting the sampling of the echo signals into precisely specified time windows. Only the signals returning from a preselected range (and thus passing through the corresponding window) are recognized. The relationship between the time window, T, with its upper and lower time limit, t_u and t_i, the length of the ultrasonic pulse overlapping the detection threshold Δt_β and the selected detection range, D, with its distance limits d_u, d_i, is given by the formulae:

$$d_u = (c/2).(t_u - t_o - \Delta t_\beta/2)$$
$$d_i = (c/2).(t_1 - t_o + \Delta t_\beta/2)$$
$$D = d_u - d_1 \qquad (7.5)$$

where c is the velocity of sound in air.

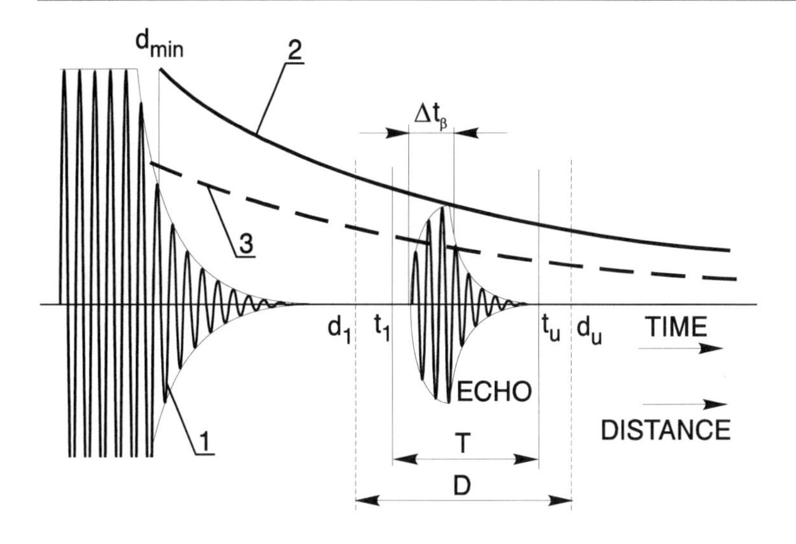

Fig. 7.9 ***The delay of the reflected signals detected by an ultrasonic transducer 1 – sweep of the signal from the transducer, 2 – expected level of the reflected signal, 3 – detection threshold, t_o – centre of the transmitted pulse, d_{min} – minimum distance for detection, T – time window with the upper and the lower limits, t_u and t_1, t_β – a part of the reflected signal overshooting the threshold level, D – selected detection range, determined by the time window, T, and by the lower, and the upper limits of distance, d_i and d_u***

Since Δt_β depends on the detection threshold, the selected detection range, D, will also depend on the amplitude of the echo. Although the intensity of the emitted ultrasound as well as the sensitivity of the receiver falls with increasing angular deviation from the main axis, some echoes from objects in other directions can also be detected, provided their reflected signals exceed the detection threshold. Highly-reflecting objects in lateral positions are likely to be detected from larger distances than low-reflectance ones. In Fig. 7.9, a diagram of the signals received as a function of time is shown, or as a function of distance from the reflecting object.

An ultrasonic transducer converts part of the electrical energy consumed, P_o, with a certain efficiency, η, into acoustic energy, P_{oa}, which afterwards is dispersed as ultrasound. When the ultrasonic beam is reflected from an object, the transducer receives the acoustic signal, P_{ia}, which this time is converted back to electrical energy, P_i, with the same efficiency, η. The ratio of the electrical signals, P_i/P_o, measured by two identical transducers which are mounted very close to each other, is called the insertion loss (IL) of the transducers. It is usually measured as the ratio of an open circuit reception voltage, E_i, to the transmission voltage, E_o (provided the transducers are identical). The logarithm of this ratio, expressed in decibels, must be reduced by 6 dB, to account for the fact that an open circuit receiving transducer (i.e. unloaded) produces twice as high an output voltage, E_o, than the same transducer with perfect impedance matching (i.e. in the state when it converts the acoustic energy to the maximum possible electrical energy P_i). Thus

$$IL = 20 \log(E_i/E_o) - 6 \quad [dB] \qquad (7.6).$$

The reflectivity, R, of the object represents the ratio of the echo intensity of a real object to the intensity of echo of an ideal object at an equal distance. R depends on the coefficient of reflection from the object's surface, and on its shape and size.

As approximations to ideal objects, sufficiently large, smooth boards can be considered, set perpendicularly to the direction of wave propagation. During the time of pulse propagation, the intensity of radiation decreases approximately quadratically with distance, due to divergence and to the exponential attenuation in air (curve 2 in Fig. 7.9).

The intensity of the echo returning from an ideal reflector is equal to the intensity of direct radiation measured at twice the distance sensor-reflector. The maximum detectable range is represented by a distance beyond which the acoustic intensity decreases to a level at which a signal returning from an ideal object is just above the noise level of the receiving amplifier. This condition must also be satisfied by low-level signals, returning after reflection from non-ideal objects. Naturally, for non-ideal objects, the detection range corresponding to a low value of the reflection coefficient is shorter than the maximum detection range. Therefore the value of the angle beyond which the interference increases is thus lowered by highly directional transducers. At higher frequencies, the spectral density of most sources of noise is low, and the high attenuation of air enables only the closest sources of noise to cause interference.

A suitable ultrasonic transducer must satisfy certain (somewhat contradictory) requirements:

- a close-covered, large radiation surface, so as to reduce sensitivity to dust and dirt (an inevitable condition of industrial use)
- high power efficiency at high levels of emitted energy
- high directivity
- cheap production (low price) of the transducer should be achieved by correct selection of the design principle.

In spite of their excellent electroacoustic properties, electrostatic foil transducers, as used for instance in self-focusing cameras, cannot be applied as sensors in industry because of their high sensitivity to dust and dirt, and their low mechanical strength. The simple piezoceramic vibrators radiate with a very low radiation power efficiency, β, in air

$$\beta = 8.Z_a/Z_c \qquad (7.7)$$

which follows from the imperfect coupling caused by the high acoustic impedance of piezoceramic materials, $Z_c = 2.5\times10^7$ kg.m^{-2}.s^{-1}, compared with the extraordinary low acoustic impedance of air, $Z_a = 400$ kg.m^{-2}.s^{-1} (recall that the acoustic impedance, Z, is the density of a material, ρ, multiplied by the sound velocity, c). Since the efficiency, η, is given by the radiated power and the quality factor, Q, of the transducer,

$$\eta = \beta Q_t \qquad (7.8),$$

solid piezoceramic vibrators can reach high power efficiency only with a high Q-factor, which is unfortunately accompanied by low spatial resolution and a long dead zone. The radiation power efficiency can be remarkably improved by using a quarter-wavelength matching layer from a material with a lower acoustic impedance. An extraordinarily stiff material was discovered by systematic research. This composite material is composed of microscopic hollow spheres of glass (diameter d = 50 μm), with epoxy resin as a base material. This coupling medium has an acoustic impedance of $Z_m = 1.3\times10^6$ kg.m^{-2}.s^{-1} ($C_m = 2600$ m.s^{-1}; $\rho_m = 500$ kg.m^{-3}). For manufacturing cheap transducers with a high directivity, disc-shaped piezoceramic vibrators are used which operate at the radial resonance frequency. An advantage of radial vibrations is that only a small amount of material is necessary to create a relatively large surface which may be excited at resonance. A further advantage of radial resonance is that all other interfering resonance modes are at sufficiently removed frequencies.

A typical example of the transducers described above is a probe with the following properties: the total diameter is 52 mm, which represents about 15 wavelengths in air. It is driven by a round disc manufactured from the piezoceramic material VIBRIT 420 R, which at resonance works in a radial mode. The coaxial ring is produced from aluminium (Fig. 7.10). The electrical power of 10 W, exciting the transducer, creates an efficient acoustical pressure of 65 Pa at a distance of 1 m. Because the efficiency at this power value is considerably lower than that of low-

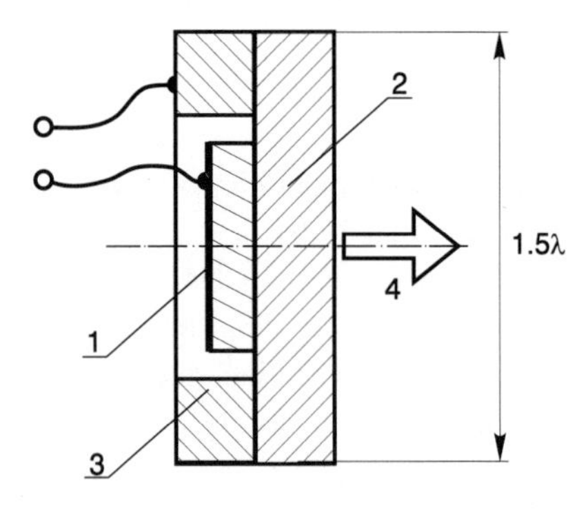

Fig. 7.10 ***Highly directionally-oriented ultrasonic transducer, 1 – piezoceramic disc, 2 – coupling (matching) layer, 3 – metallic ring, 4 – direction of maximum radiation***

power operation, it is not reasonable to increase the power above the value P_{max} = 10 W, as the acoustic power produced begins to saturate.

The maximum power, P_{max}, usable for the transducers, is limited by their maximum capacity for concentration of energy, because thermal effects are negligible in pulse-echo systems. Therefore the maximum power is proportional to the volume O of the transducer, and to the operational frequency f. Since the transducers are usually adjusted to resonance at their working frequency, their volume is also inversely proportional to the third power of the working frequency, which in consequence results in a decrease of the maximum usable power, proportional to the square of frequency. Thus

$$\left.\begin{array}{r} P_{max} \approx O.f \\ O \approx f^{-3} \end{array}\right\} \Rightarrow P_{max} \approx f^{-2} \; .$$

This means, for example, that the maximum allowable power that may be applied to a 400 kHz transducer is 1/25 of the allowable power for the 80 kHz transducer.

In Fig. 7.11 examples of typical applications of ultrasonic proximity sensors are shown.

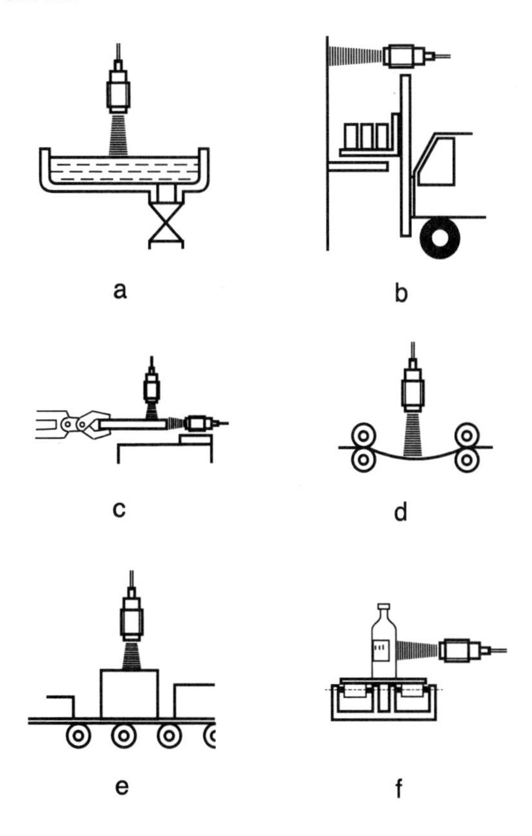

Fig. 7.11 ***Examples of application of an ultrasonic proximity sensor***

While proximity sensors have, as a rule, a limited service range up to about 6 m, a relatively high operational frequency, from 80 to 400 kHz, and small dimensions, ϕ 30 × 95 mm, or ϕ 75 × 115 mm, ultrasonic level-gauges are characterized by rugged construction, a working frequency between 20-40 kHz, dimensions of ϕ 285 × 180 mm and a service range up to 60 m.

7.6 Ultrasonic flow gauges

In section 4.4 the principles of ultrasonic flow-gauges were introduced. One of the principal advantages of this type of flow-gauge is the ability to measuring flow without intruding into the pipeline. Therefore in this section attention is devoted to this type of flow measuring instrumentation in more detail.

In Fig. 7.12, two principles of fastening probes onto a pipe are introduced. In case a, one of the probes is positioned above, and the other one below the pipe. In case b, both the probes are on one side of the pipe, whereby those ultrasonic waves which were reflected from the inner pipe wall arrive at the receiving probe.

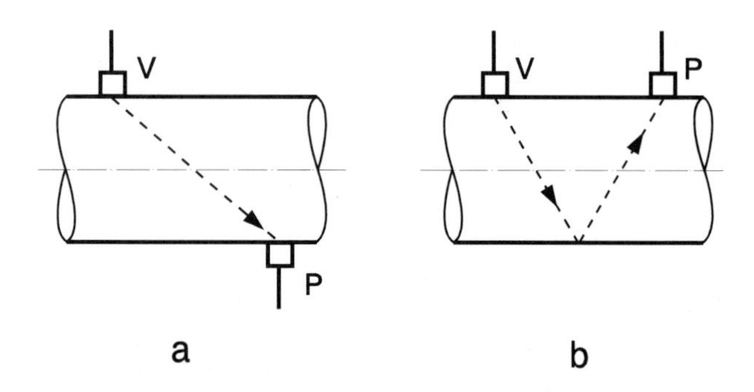

Fig. 7.12 ***Examples of mounting a probe on the outer side of a pipeline, a – probes above and below the pipe, b – probes at one side of the pipe***

As mentioned before, the method of determination of the flow velocity is based on measurement of the time, T_1, of transition of the emitted pulse from the upper side of the stream to the lower one, and the time, T_2, of transition from the lower side of the stream to the upper one. These time signals, after appropriate processing, are converted to electrical pulses. The time intervals, T_1 and T_2, can be determined as a sum of times t_1 and t_2, during which the ultrasonic pulse propagates through the liquid, and a time τ, when it propagates through an extension, and the wall of the pipe (Fig. 7.13)

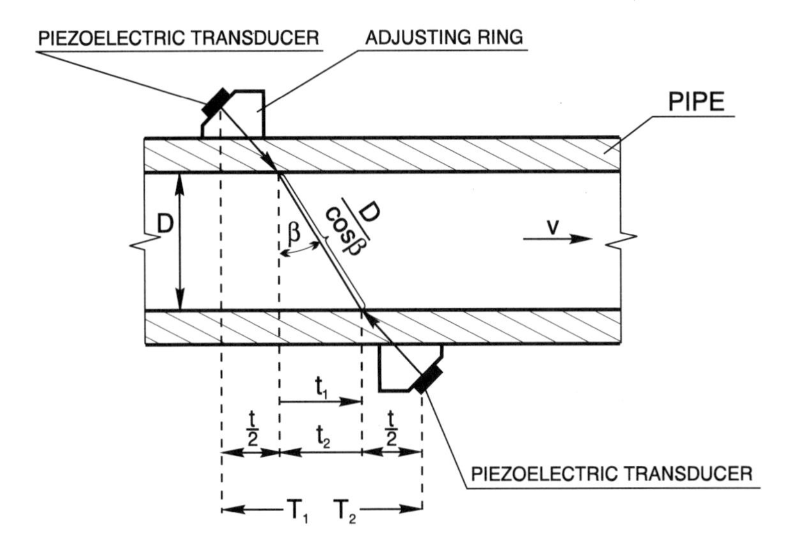

Fig. 7.13 ***The principle of ultrasonic flow measurement by probes mounted on the outer surface of a pipe***

$$T_1 = t_1 + \tau \qquad T_2 = t_2 + \tau \tag{7.9}$$

In general, the transition time of the ultrasonic wave motion in both directions through the flowing liquid can be written as

$$t_1 = \frac{D/\sin\beta}{c + v.\cos\beta} \qquad t_2 = \frac{D/\sin\beta}{c - v.\cos\beta} \tag{7.10}$$

where D - inner pipe diameter

β - vertical angle of propagation of the ultrasonic wave

c - velocity of sound in liquid

v - velocity of the liquid flow.

In equation (7.10), the numerator $D/\cos\beta$ characterizes the distance which the ultrasonic wave has to travel in a direction at an angle β relative to the main axis (direction of the flow), and the term $v.\sin\beta$, in the denominator, characterizes the term of the flow velocity vector in the direction of propagation of the ultrasonic wave. The change of the propagation velocity of an ultrasonic pulse at a slanted trajectory (at constant velocity of propagation, c, in a stationary liquid medium) depends only on the flow velocity. Therefore, the time difference, ΔT, of the times T_1 and T_2 is proportional to the flow velocity, thus

$$\Delta T = T_2 - T_1 = t_2 - t_1 = \frac{2D.\sin\beta}{c^2.\cos\beta}.v \tag{7.11}$$

This expression results from the condition that the term $v^2.\sin^2\beta$ is so small compared to c^2 that it can be neglected.

If the time of propagation of an ultrasonic wave in a stationary liquid medium is denoted as T_o, then

$$T_o = T_1 = T_2 = \frac{D/\cos\beta}{c} + \tau \qquad (7.12),$$

hence

$$c = \frac{D/\cos\beta}{T_0 - \tau} \qquad (7.13).$$

These equations enable one to write equation (7.11) in the form

$$\Delta T = \frac{\sin 2\beta}{D}\,(T_o - \tau)^2.v \tag{7.14}$$

where all terms are either constants or measured quantities.

The flow velocity measured by an ultrasonic flow-gauge is an average linear velocity, whose vector is at a certain angle relative to the velocity vector of the ultrasonic waves which propagate in a radial direction. Of course, the liquid flow velocity in a cylindrical pipe has, in general, a non-linear character because the velocity is higher in the middle of the pipe. Therefore, following determination of the average velocity, v_p, of the flowing liquid, a correction should be made for the velocity distribution across the pipe.

If the correction coefficient is denoted as $K = v/v_p$, then in the range of velocities where the theory of turbulent flow can be applied, the K value can be enumerated as

$$K = 1.119 - 0.011.\log\{Re\} \tag{7.15}$$

where Re is the Reynolds number for the flow.

The total flow is determined as the product of the cross-sectional area and the mean flow velocity, i.e.

$$Q = \frac{\pi D^2}{4} \cdot v_p = \frac{1}{K} \cdot \frac{\pi D^2}{4} \cdot v = \frac{1}{K} \cdot \frac{\pi D^2}{4} \cdot \frac{D}{\sin 2\beta} \cdot \frac{\Delta T}{(T_0 - \tau)^2} \tag{7.16}$$

where T_o is the time of propagation of an ultrasonic wave in a stationary liquid. In the case of a flowing liquid, this time can be evaluated approximately as $T_o = (T_1 + T_2)/2$. The time, τ, corresponds to the time of propagation of an ultrasonic wave outside the liquid, through the pipe walls and/or the extension. For this reason, it is essential that one must determine precisely the pipe wall thickness and the velocity of sound within the material of the pipe.

When using flow-gauges with external fastening of probes, the flow is calculated from the measured values of the transition time of an ultrasonic wave between two probes, the calculated values of the pipe, the tilt angle of the wave trajectory relative to the flow axis, the time of transition of the wave through the pipes and the extension thickness.

One of the principal functional features of the probes, and the associated electronic circuits, is their resolution of the time measurement. It should be noted, for instance, that for a pipe with an inner diameter of 25 mm, the time difference corresponding to a liquid flow velocity of 1 m/s is only 10 ns. Therefore a flow-gauge with an accuracy of 1 % must have a time resolution of at least 0.1 ns.

Modern ultrasonic flow-gauges utilize time-digital processing (TDP) for measuring velocity. A typical block diagram is shown in Fig. 7.14. The circuit works in the following way:

1) The probe, 2, emits an ultrasonic pulse after the arrival of a signal from the central processor, 8, via circuit 4, which is synchronized in phase with the oscillations of the voltage controlled generator, 9.

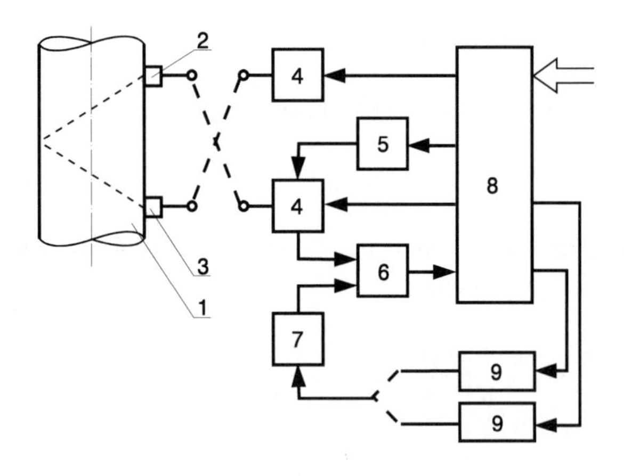

Fig. 7.14 ***A block diagram for digital processing of a flow measurement, 1 – pipe, 2 – probe I, 3 – probe II, 4 – receivers-transmitters, 5 – digital-to-analog converter, 6 – time detector of wave propagation, 7 – counter, 8 – central processor, 9 – voltage controlled oscillator***

2) The ultrasonic pulse is detected by probe 3 after transition through the pipe and is amplified. The amplification coefficient is controlled by the output voltage of the digital-to-analog converter, 5.
3) The digital-to-analog converter, 5, operates with five different voltage signals from the central processor, 8. Two voltages are used for control of the pulses from the detected signals, which propagate in the direct and in the antiparallel directions. The other three control voltages are used for distinguishing the directions of the detected signals, for detection of the measuring point at the instant of detection and for detection of the noise level.

4) The circuit of the time detector, 6, controls the voltage controlled oscillator, 9, in order that the instant of detection was coherent with the phase of the oscillations of the voltage controlled oscillator. This circuit measures the number of time marks from the instant of emission till detection of the signal.

The input information about the technical characteristics of the pipe is processed via the interface of the flow gauge, in the manner of a dialogue. The interface determines the measuring conditions, and converts the measured data to values of the expected physical quantities. The interface works in the following way:

1) following connection to the mains, the display is ready to receive entries
2) entering and recording of data about the technical characteristics of the pipeline via keyboard, such as the outer diameter, pipe material, pipe wall thickness, via a dialogue
3) depiction of the data on the display
4) transfer of the data into the TDP block - as instructions about the synchronization mode, change in the flow velocity, etc.
5) calculation of the flow velocity values - indication of the normal and the deviated values, and the values v and T_o.
6) processing of the data - calculation and conversion of the measured values to the necessary operating values (e.g. instantaneous flow, integrated flow), processing of the input data about the measurement conditions via keyboard
7) output of data - depiction on the display, analog output, output to printer, etc.

The majority of ultrasonic flow-gauges produced by various manufacturers [7.27 - 7.29], work on this principle, with only minor deviations.

In section 4.4, mostly the measurement of liquid flow was discussed. In a similar manner, gas flow can also be measured. With regard to different acoustic impedances of solid materials and gases varying by 6 orders of magnitude, built-in probes inside the pipes must be used. Several manufacturers deal with the development and manufacture of such instruments [7.30, 7.31]. A typical representative of this group is instrument VMA 1 of the SICK company, for measuring the flow of smoke gas in stacks.

The principle of measurement by ultrasound enables contactless integral measurement (across the whole cross-section of the channel). The average flow velocity is measured with very high accuracy, and it is possible to distinguish even the smallest differences of velocity. The instrument has neither built-in components inside the stack nor parts that wear out so that it needs practically no service.

The ultrasonic transducers are mounted on flanges, diagonally, on the outer walls of the stack. The measured signals are processed and evaluated in the appropriate electronic circuitry (Fig. 7.15).

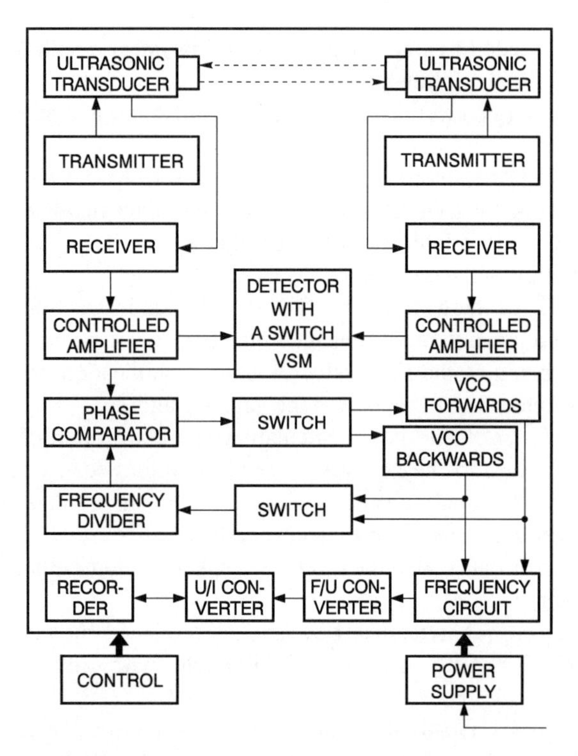

Fig. 7.15 ***Block diagram of ultrasonic flow meter VMA 1***

The function and the time development of the measurement is controlled by a crystal oscillator. After division of the selected oscillator frequency, the pulses are led alternately to the ultrasonic transmitters. After their reception, there follows a conversion to electrical pulses, which are further processed into the form of data. As the transmitter output and the receiver input are connected, the receiver input must be blocked during transmission so that saturation does not arise. The signal, which appears at the receiver with a certain delay, is amplified and led to the control amplifier. Then it is rectified, shaped via a Schmitt trigger and the output pulse is led to a phase comparator.

Let us observe a signal which is emitted by the transmitter, V, and received by a receiver, P, after the signal has arrived at the phase comparator input, E1.

The phase comparator output is connected to a voltage controlled oscillator, VCO, which oscillates at a frequency of about 500 kHz. The frequency of the VCO is divided by a frequency divider until the leading edge of the 500 kHz VCO pulse corresponds with the edge of the received signal. The divided signal is led to the input E2, of the phase comparator. Now the phase of the received measuring signal is compared with the signal from the VCO oscillator. If the phases do not correspond with each other, the VCO oscillator is finely adjusted (tuned in). Then the control circuit switches to the reverse process, i.e. with the transmitter P and the receiver V.

Now the ultrasonic wave propagates through the gas in the opposite direction. The signal processing procedure described above is repeated, the received signal is compared with the divided VCO frequency, P, of the oscillator, in the phase comparator. Now both the frequencies from the VCO oscillators are read. The frequency difference is zero, as the frequency of both the VCO oscillators is the same (the flow velocity is $v = 0$ m/s).

If the flow velocity is non-zero, $v > 0$ m/s, the time of transition of ultrasonic pulses in the direction of the gas flow shortens, and, vice versa, it lengthens in the antiparallel (opposite) direction.

This results in a difference in the frequencies of both the VCO oscillators. Now, the frequencies have changed to such an extent that, after division, they correspond again with the frequency of the received signal. If these frequencies are subtracted one from the other, the result of subtraction is a difference frequency, which is directly proportional to the velocity of the gas flow. The difference frequency is converted to a voltage and then to a current, which is recorded.

This instrument, with regard to its design and its parameters, is appointed specially for measuring the velocity of gas in chimneys and stacks. The (inner) stack diameter can be in the range from 0.4 to 10 m. The measuring range of the instrument is 0.1 - 50 m/s. The analog output to the recorder is 0 - 20 mA, or 4 - 20 mA. Its accuracy is 5 % of the maximum value.

7.7 Ultrasonic defectoscopy (NDT)

In Chapter 5, the fundamental principles of NDT (Non Destructive Testing) were described. In this section, some practical applications of NDT, and broader perspectives, will be mentioned.

As already mentioned, transmission methods (based on the transition of the beam through the medium) are no longer used in classical defectoscopy. In this section, one non-specific (although very interesting) application will be described, for detection of bubbles and inhomogeneities in chipboards, flaxboards and laminated sheets [7.32].

Trienco, Model 506 - a system for ensuring quality - is an ultrasonic instrument capable of detecting bubbles and failures in panels at normal manufacturing speeds. The check is performed without physical contact with the product. The output data are used for automatic control, marking the products, for auxiliary instruments which give warning signals, and for the purposes of collecting data. The Model 506 is equipped with independent subsystems, which are connected with the central measuring system. Each of these subsystems consists of a transmitting transducer, a receiving transducer, a module for processing the signals and a relay. The concept of using independent modular subsystems increases the overall reliability and operational ability of the system. If a failure occurs, all of the modules can easily be exchanged and repaired.

The transmitting transducer, placed beneath the measurement volume, radiates sonic waves of high frequency which are directed upwards, to the bottom of the panel, or other product under test. Part of the emitted sound penetrates into the panel (plank board, etc.), passes through it and surfaces on the opposite side, where it is detected and converted to an electrical signal by means of a receiving transducer. A bubble or a failure which is found by the beam along its travel causes a change in the beam propagation. The electrical signals produced by the receiving transducers indicate the internal integrity of the product. These signals are sent to a processor module in which the values are interpreted in an electronic form and the processor determines whether the measured spot is good or bad, and if a failure was found, whether its size is still acceptable or if it exceeds the preset limits allowable for an undamaged product. Every emitting transducer generates a limited beam with a diameter of about 4.5 mm. During the test, the panel continually moves past the transducers, and each transducer emits about 60 ultrasonic pulses per second, which effectively irradiate a 45 mm wide surface along the whole length of the panel. In this way, the existence of bad sections of the panel is indicated.

The classical ultrasonic pulse method is a basic element of ultrasonic defectoscopy. Pulse systems are suitable for the location of a defect, evaluation of its size and, last but not least, for digital processing of the test results. This method substantially improves the automated inspection process, and in some cases of complex multiple-probe systems, it is an inevitable requirement for proper evaluation of the test. Without a digital evaluation, the automatic inspection of large complexes would be impossible, such as the bodies of pressure vessels for nuclear power plants, testing of metallurgical products in the course of production, etc. The large amount of information obtained in a fraction of a second cannot be processed in other than a digital way. Only a computer can evaluate the results of such a test because it can simultaneously compare the current results with the previous ones.

In order that the productivity of the inspection corresponds to the actual requirements, ultrasonic equipment must satisfy the following requirements:

- automated remote or programmable control of drives which control the movement of the probe system
- automated data collection
- automated data processing
- automated calibration and testing of the operation of the equipment
- automated search for the zones to be examined
- computer controlled pre-setting and compensation for the influence of the distance of the defects.

Usually, a common projection, A, is utilized in pulse methods, and only two parameters are used:

- the echo height
- the echo transition time

In modern instruments, these parameters are digitized [7.33 - 7.34], processed by a dedicated microprocessor and projected in a suitable pattern which enables one to obtain new information about the defect. The possibility of storing information about the defect allows later comparison with newer data. The microprocessor controls all phases of processing and evaluation of the echo.

The demands, in practice, of automated testing of moving objects, leads to the requirement of contactless generation and detection of ultrasound.

With the aid of picosecond laser pulses, it is possible to generate pulses of acoustic energy which are sufficiently short (~0.5 ns) to allow measurement of polymer layers with a thickness of about 16 μm. Besides this, the wide-band character of acoustic pulses also enables measurement of attenuation from 10 MHz up to 160 MHz.

It is also possible to use laser-induced generation of ultrasound for the examination and study of microstructural parameters. A laser interferometer as a detector, and its associated remote operation system, is able to distinguish among various kinds of steel with different grain size, by comparing the frequency spectrum of the ultrasound received.

A laser-induced ultrasonic system enables one to test samples of hot steel directly during the process of continuous casting. Measurements on steel at temperatures up to 1200 °C were carried out, on rough surfaces and in a dusty and turbulent air flow.

In order to ensure the safety of operation in a nuclear power plant, it is necessary to check periodically the state of the material of the nuclear reactor pressure vessel. The goal of the regular service inspections is to discover any possible internal material defects and to estimate their seriousness. The dynamical operational strain of a body gives rise to fatigue ruptures and fissures which are proportional to the direction of the main mechanical strain, or else they traverse the vessel body in the welding planes. Thus the likely direction of the ruptures are oriented in the planes perpendicular to the surface, or slightly deflected from this direction.

In operational tests of pressure vessels by ultrasound, besides fatigue ruptures oriented perpendicularly to the surface, less frequently occurring defects are sought for, such as possible separation of the protective stainless steel layer from the basic material of the reactor vessel.

In testing the body of a pressure vessel where the wall thickness is around 200 mm, several pairs of ultrasonic angular probes are used in a tandem arrangement, in such a way that it is possible to check through the entire wall thickness while the set of probes remains in the same position. The measurement is divided into several intervals (segments), each of them having a different examination depth, with some mean depth.

The depth segments into which the whole wall thickness is divided are tested in separate time intervals of the measurement cycle. An arrangement of nine probes is shown in Fig. 7.16, as an example of one possible setup. Of these, probes 1 to 8 emit pulses in eight successive time intervals, and probe 9 serves as a receiver. The number of examined depth segments corresponds to the number of probes.

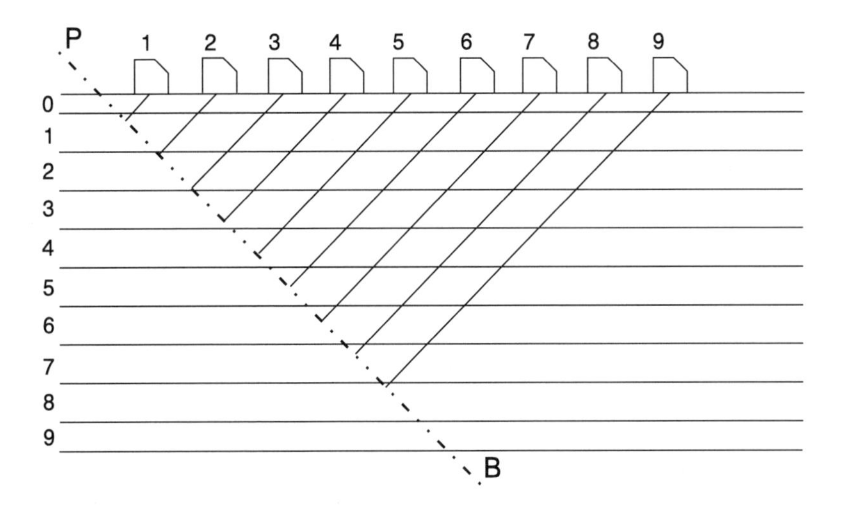

Fig. 7.16 ***A setup of probes for detection of defects in the direction perpendicular to the surface, by an indirect reflection along the connecting line P - B***

The pressure vessel can be examined by ultrasound also from within. In the first case, the ultrasonic probes are fastened by gimbal suspensions on a tray which moves on rails and scans the outer surface of the pressure vessel. A more suitable, and therefore a more widely used testing mode is that from within the vessel. An immersion-type acoustic coupling is used with advantage in this case, because the vessel is filled with water.

In examination from within the vessel, manipulator equipment is inserted into the vessel, which bears the probe setup. The probe holder itself can be tilted at various angles, so as to enable testing of welds in various parts of the pressure vessel, for example, its bottom section, etc.

The electroacoustic transducers inside the probes are connected to their generators and gated preamplifiers, which are positioned close to the probes. The preamplifiers are gated open only in the time interval when the respective probe transmits and receives. They are arranged in two groups, each of them having a common output connected to the input of one of two amplifiers with a controlled amplification. Their output is common again, and it is connected to the input of a logarithmic amplifier. The signal from the output of the logarithmic amplifier passes through a detection circuit, in which videosignals are created with an amplitude of the order of volts.

In the sampling amplifier, signals are obtained which are suitable for analog-to-digital conversion. The output, in the form of digitally coded, 6-bit signals passes, via buffer storage, either directly to a computer or else is temporarily recorded.

Digital data on the echo amplitudes caused by the defects, and those from the opposite surface of the pressure vessel, together with the digital data and the coordinates of geometrical position, are processed by a computer in an on-line or off-line manner.

7.8 References

7.1 *Commercial press:* SAC graph/pen. Informative prospectus, SAC Comp.

7.2 *Commercial press:* Graph/pen briefs No. 1, May 1973. Issued SAC Comp.

7.3 *Commercial press:* Digitizing with a difference. Prospectus of SAC (Science Accessories Corporation), 200 Watson Boulevard, P.O. Box 587 Stratford, CT 06497, USA

7.4 *Commercial press:* Model GP-9. Two-dimensional Sonic Digitizer, Prospectus, SAC Comp.

7.5 *Commercial press:* Model GP-12. 3D Digitizer. Prospectus, SAC Comp.

7.6 *Driels M.R., Pathre U.S.:* Significance of Observation Strategy on Design of Robot Calibration Experiments, J.Rob.Systems 7, 1990, No. 2, 197-223

7.7 *Luck R., Ray A.:* Failure Detection and Isolation of Ultrasonic Range Sensors for Robotic Applications. IEEE Transactions on Systems, Man and Cybernetics 21, 1991, 221-227

7.8 *Marioli D., Sardini E., Taroni A.:* Ultrasonic Distance Measurement for Linear and Angular Position Control. IEEE Trans. on Instrum. and Measurement 37, 1988, No. 4, 578

7.9 *Kweon I., Kuno Y.,Watanabe M., Onoguchi K.:* Behavior-Based Mobile Robot Using Active Sensor Fusion. Proc. IEEE Int. Conf. on Robotics and Automation, IEEE Comp. Soc. Press, Los Alamitos, Vol. 2, 1992, 1675-82

7.10 *Obaidat M.S., Abu-Saymeh D.S.:* Methodologies for Characterizing Ultrasonic Transducers Using Neural Network and Pattern Recognition Techniques, IEEE Transaction on Industrial Electronics, Vol. 39, 1992, Iss. 6, 529-36

7.11 *Shetty K.B., Ranganathan R., Chidambara M.R.:* Ultrasonic Sensor Interface to an Articulated Joint Robot for Object Detection and Distance Measurement, Proc. SPIE (The International Society for Optical Engineering), Vol. 1571, 1991, 114-119

7.12 *Shikomohbe A., Ma S.:* Ultrasonic Measurement of Three-Dimensional Coordinate. In: Proc. 11th World Congress of IMEKO, IMEKO Press, Houston, 1988, 441-447

7.13 *Warnecke H.J., Schiele G.:* Performance Characteristics and Performance Testing of Industrial Robots-State of the Art. In: Proc. Robotic Europe Conf, 1984, Springer Berlin

7.14 *Prenninger J.P.:* Contactless Position and Orientation Measurement of Robot End-effectors. Proc. IEEE Int. Conf. on Robotics and Automation, Atlanta, May 1993, IEEE Comput. Soc. Press, Los Alamitos, Vol. 1, 1993,180

7.15 *Commercial press:* Weather Measure Corp.,Sacramento,USA: W 115 Sonic Anemometer-Thermometer. Scientific Instruments and Systems Catalogue, 1078

7.16 *Commercial press:* Kaijo Denki Co., Ltd, Japan: Digitized Ultrasonic Anemometer-Thermometer, model DAT-300 (technical description)

7.17 *Kéry M.:* Spôsob číslicového merania priestorových zložiek vektora okamžitej rýchlosti vetra a zariadenie na jeho prevádzanie. Patent (Author's Certificate), ČSSR, No. 204 166 (in Slovak)[1]

7.18 *Commercial press:* Ultrasonic Precision Thickness Gauge CL 304, Wells Krautkrammer Blackhurse Road, Letchworth, Herfordshire SG6, G.B.

7.19 *Commercial press:* Precision Ultrasonic Thickness Gauges, Ultrasonics 29, 1991, No. 2, 181-182

7.20 *Commercial press:* New Principle for Digital Thickness Gauge, Ultrasonics 22, 1984, No. 4, 150

7.21 *Tsutsumi M., Ito Y., Masuko M.:* Ultrasonic In-Process Measurement of Silicon Wafer Thickness, Precision Engineering 4, 1982, 195-199

7.22 *Commercial press:* Ultraschall-Abstandssensoren, Honeywell AG, Kaiserleistrasse 39, W-6050, Offenbach, FRG

7.23 *Commercial press:* NIVELCO KFT, Béke n.118, 1131 Budapest, Hungary

7.24 *Commercial press:* Microsonic Sensor SM 140/190, Hyde Park Electronics Inc., 4547 Gateway Circle, Dayton, OH 45440, USA

7.25 *Commercial press:* Migatron RPS 400, Migatron Corp., 120E Burlington, La Grange, IL 60525, USA

7.26 *Commercial press:* Sonar - BERO, Siemens Aktiengesselschaft, D-8520, Erlangen, FRG

[1]A method of digital measurement of the spatial terms of an instantaneous velocity vector for wind and a device for its implementation

7.27 *Commercial press:* Microprocessor-Based Portable Flowmeter, Ultrasonics 25, 1987, No. 6, 385-386

7.28 *Commercial press:* New Ultrasonic Flowmeter, Badger Meter Inc., 6116 East 15th Street Tulsa, Oklahoma, 74155, USA

7.29 *Commercial press:* Ultraschall-Durchflussmesser für Flüssigkeiten, Eduard Schinzel GmbH, Studenygasse 16, A-1110 Wien

7.30 *Commercial press:* British Gas Ultrasonic Flowmeter, Ultrasonics 25, 1987, No. 2, 125

7.31 *Commercial press:* VELOS 500 - Gas Velocity Monitor, Erwin Sick GmbH, Nimburger Strasse 11, D-7801 Reute, FRG

7.32 *Commercial press:* Model 506 Quality Assurance System, McCarthy Products Comp., Seattle, WA 98115, USA

7.33 *Obraz J.:* Zkoušení materiálu ultrazvukem, SNTL, Prague, 1989 (in Czech)[2]

7.34 *Commercial press:* Automatische Ultraschall-Prüfanlagen, Karl Deutsch, Prüf und Messgerätebau, D-5600, Wuppertal, FRG

7.35 *Každys R. J.:* Ultrasonic Information Measuring Systems, "MOSKLAS", Vilnius, 1986, Latvia

7.36 *Obraz J.:* Použití mikroprocesoru při automatizaci ultrazvukového zkoušení. Proceedings, Defektoskopie 88, DT ČSVTS, Bratislava, 1988 (In Czech)[3]

[2]Testing of materials by ultrasound

[3]Use of a microprocessor for the automation of ultrasonic testing

Index